THE MÖSSBAUER EFFECT AND ITS APPLICATIONS IN CHEMISTRY

ÉFFEKT MËSSBAUÉRA I EGO PRIMENENIYA V KHIMII

ЭФФЕКТ МЁССБАУЭРА И ЕГО ПРИМЕНЕНИЯ В ХИМИИ

THE MÖSSBAUER EFFECT AND ITS APPLICATIONS IN CHEMISTRY

by

V. I. GOL'DANSKII

Corresponding Member
of the Academy of Sciences of the USSR

Authorized Translation from the Russian

CONSULTANTS BUREAU
NEW YORK
1964

The original Russian text, published for the Institute of Chemical Physics by the Academy of Sciences Press in Moscow in 1963, has been supplemented by the author for the English edition.

Виталий Иосифович Гольданский

Эффект Мёссбауэра и его применение в химии

Library of Congress Catalog Card Number 64-21682

ISBN-13: 978-1-4684-1556-8 e-ISBN-13: 978-1-4684-1554-4
DOI: 10.1007/ 978-1-4684-1554-4

PREFACE

The Mössbauer effect is not only one of the simplest and most beautiful discoveries of nuclear physics, it is one of those discoveries which, to the highest degree, tend to produce and develop new ties between nuclear physics and the other branches of physics as well as the other natural sciences and engineering. Two of our earlier papers which form the basis of this monograph were devoted to "propaganda" for the Mössbauer effect among chemists and "propaganda" for the chemical applications of the effect among nuclear physicists. The first paper, presented in January 1962 at the Scientific Colloquium of the Learned Council of the Theory of Chemical Structure, Kinetics, and Reactivity of the Division of Chemical Sciences of the Academy of Sciences of the USSR, gave a general description of the Mössbauer effect and the methods by which it is observed, and enumerated a number of problems in structural chemistry, chemical kinetics, and radiation chemistry that might well be studied by observing the Mössbauer spectra, in particular those of organotin compounds.

The second paper, presented in July 1962 at the working conference on the Mössbauer effect at the Joint Institute of Nuclear Studies, was entirely devoted to the many possible chemical applications of this new phenomenon, the most interesting of which seem to pertain particularly to the chemistry of complex and metalloorganic compounds. In preparing this paper for the printed transactions of the Dubna conference, a detailed analysis of our own and foreign published results on chemical applications of the Mössbauer effect to Fe and Sn nuclei was added. The rate at which new data on the Mössbauer effect are appearing is exceedingly rapid so that it would be premature to attempt a summary of the work done in this direction, the more so since, as with every other method of physicochemical investigation, observation of Mössbauer spectra alone is insufficient and comparison with the many other kinds of data that are now becoming available is required if conclusive information on the structure of molecules is to be obtained.

However, the principal problem of this monograph, conceived as a review addressed primarily to chemists, but greatly exceeding the length acceptable to the journal "Uspekhi khimii" (Advances in Chemistry), lies not so much in presenting unquestionable and final results, as in stimulating further studies and outlining the results that may be expected from them, and in creating the widest possible interest in all kinds of chemical applications of the Mössbauer effect. A perusal of recent publications shows that this emphasis is placed quite correctly. Both in the Soviet Union and abroad, papers on the use of the Mössbauer effect in chemistry for the most part deal only with increasing the number of examples and illustrations of successful solutions, and not with a fundamental broadening of the scope of the method.

The reader will certainly find that this monograph is not without omissions and defects, and comments concerning these will be received with thanks.

In conclusion, the author expresses his sincere gratitude to Yu. M. Kagan, S. V. Karyagin, E. F. Makarov, G. L. Slonimskii, V. V. Khrapov, and E. M. Shustorovich for discussing some of the questions treated in this book, and for a number of valuable comments.

V. I. Gol'danskii

CONTENTS

INTRODUCTION

Quite a broad range of phenomena that exhibit a relation between various nuclear characteristics and the structure of the electron shells surrounding the nucleus – in particular, the valence shells, which determine the chemical properties of matter – is known today. These phenomena include the following:

1. The isotope shift in optical spectra, due to a difference in Coulomb interaction energy of the optical electron with the nuclei of different isotopes of the same element, as a result of the different dimensions of the nuclei. The magnitude of this shift is [1, 2]

$$\frac{\Delta\lambda}{\lambda} \approx 10^{-6}.$$

2. The isomer shift in optical spectra, i.e., the change in wavelength of an optical spectrum line on going from the ground state to an excited (isomeric) state of the same isotope. The effect is due to the same causes as the isotope shift, but is several times weaker [3, 4].
3. The chemical shift in the nuclear magnetic resonance lines, due to difference in screening of the nucleus by the valence electrons, and – as a result of this – to the difference in effective value of the external magnetic field acting on a given nucleus in different chemical compounds [5-7].
4. The dependence of the form of the hyperfine structure of the rotational lines, or of the nuclear quadrupole resonance spectra, on the chemical bonds of the isotopes under study [5, 8-11], due to interaction between the nuclear quadrupole moment Q and the inhomogeneous electric field in the region where the nucleus is located In this case the electric field gradient $q = \partial^2 V/\partial z^2$, which gives the departure of the charge distribution around the nucleus in question from spherical symmetry, depends both on the electron shell structure of the atom itself and of the whole molecule, and on the macrostructure of the crystal.

None of the phenomena mentioned has anything to do with transformations of the nuclei in question. However, in the last 10 or 15 years, studies have also been made on some chemical changes (i.e., due to the electron shell structure of radioactive isotopes) in some of the properties of nuclear transformations. These include:

1. The chemical change in the lifetime of radioactive isotopes transforming by an electron capture mechanism [12, 13]. This is due to the fact that the electron capture rate constant is proportional to the electron density $|\psi(0)|^2$ in the region where the nucleus is located, and this density is essentially dependent on the electron shell structure. The changes of this sort that are observed are very weak ($\Delta\tau/\tau \leq 2\cdot 10^{-3}$), but there are specific conditions under which they may play an important role, as, for example, in K-capture in Be^7 nuclei when the beryllium atoms are strongly ionized in a thermonuclear fuel mixture [14].
2. The chemical change in the rate of isomeric nuclear transitions accompanied by strong internal electron conversion [15, 16] (changes have been observed up to $\Delta\tau/\tau = 2\text{-}3\cdot 10^{-3}$). The effect is dependent not only on the s-electrons, but on the state of many shells with different n and l, so that it is very difficult to interpret quantitatively.

New and rich possibilities for making a study of the effect of chemical structure on nuclear transformations and, hence, of obtaining additional information on chemical structure and on the change in chemical properties under the influence of various factors have appeared as a result of the discovery of the Mössbauer effect [17-19].

The Mössbauer effect is known as recoilless gamma resonance fluorescence (radiation, absorption, and scattering), i.e., no part of the energy is expended in recoil of the nucleus emitting or absorbing the gamma quanta.

Resonance transitions in nuclei are characterized by exceedingly high sensitivity to the smallest departures from the resonant energy. Accordingly, transfer – occurring because of the law of conservation of momentum – of part of the energy of the gamma quantum to recoil of the nucleus (like the loss of part of the energy of a shot to the recoil of the gun barrel) destroys the resonance conditions. However, in case the nucleus emitting or absorbing the gamma quantum is bound in a crystal lattice, the recoil momentum may no longer be taken by the nucleus alone, but by the lattice as a whole. In this case, the recoil energy, inversely proportional to the mass of the radiator or absorber, is incomparably less and no longer prevents resonant conditions from being set up. As a result, it becomes possible to observe nuclear resonance gamma fluorescence under conditions where it is unusually sensitive to any external factor or effect that is even to the slightest degree capable of changing the resonance energy. Such external factors are, in particular, the different chemical states of the nuclei emitting and absorbing the gamma quanta. The possibility of getting a quantitative measure of these differences from the Mössbauer effect makes it potentially very important in chemistry. In the present monograph, we shall give a short description of the Mössbauer effect and then give a more detailed discussion of the first accomplishments and the outlook for its application to chemical problems.*

* The few years since Mössbauer's discovery have seen the appearance of a large number of original papers and also a number of reviews devoted to this problem [20-38]. Since this monograph is oriented principally toward chemists, we give only a cursory account of the theory of the Mössbauer effect and its applications to physics, referring those who are interested in such questions to other reviews, in particular [20-25].

CHAPTER I

RESONANCE FLUORESCENCE

Energy and Width of Resonance Transitions

After Rayleigh had predicted and Wood in 1904 had given an experimental demonstration of the existence of fluorescence, i.e., resonance scattering and absorption of light, the discovery formed the subject of numerous successful investigations in atomic physics and optics [39]. Although Rayleigh's predictions were made on the basis of a purely classical (mechanical) description of resonance phenomena, the very existence of such phenomena in microscopic systems is of course a purely quantum property, associated with the presence of spectrum lines characteristic of transitions between definite energy levels.

Atomic nuclei, like atoms and molecules, are typical examples of microscopic quantum systems, and there was thus every reason to expect that resonance fluorescence would also be observed for the gamma rays emitted or absorbed in transitions between nuclear energy levels. Nevertheless, the search for resonance gamma fluorescence begun in 1929 [40] remained unsuccessful for more than twenty years. Then the phenomenon was finally observed, but, until the discovery of the Mössbauer effect in 1958 [17-19], only under very specific conditions (about which we shall speak below); thus the observations did not receive particularly wide publicity. What, then, is the basic difference in the conditions under which atomic (optical) and nuclear (gamma) resonance fluorescence is observed? To answer this question, we must remember first of all that any excited level is characterized both by the resonance excitation energy E_r and by its natural width Γ. This width, which is related to the mean lifetime τ of the excited state by the indeterminacy relation $\Gamma\tau = \hbar = 1.05 \cdot 10^{-27}$ erg · sec, defines the "tuning accuracy" required for going into resonance. If the excited state can decay in several different ways (first, second, third, etc.), the total excited level width Γ is equal to the sum of all the partial widths,

$$\Gamma = \Gamma_1 + \Gamma_2 + \Gamma_3 + \cdots = \sum_i \Gamma_i,$$

and since each of the partial widths Γ_i is related to the mean lifetime τ of the system with respect to a given type of decay by the equation $\Gamma_i \tau_i = \hbar$, as the number of types of decay increases, the mean lifetime of the excited state drops:

$$\frac{1}{\tau} = \frac{1}{\tau_1} + \frac{1}{\tau_2} + \frac{1}{\tau_3} + \cdots = \sum_i \frac{1}{\tau_i}$$

We shall consider a system with a single ("insulated") excited level and take the excitation probability of the level to be unity when exactly the resonance energy E_r is added to the system. Then the resonance excitation probability for any other energy E (i.e., the form of the resonance line) is given by the Breit–Wigner dispersion formula [41] and is equal to

$$W(E) = \frac{1}{1 + \left(\frac{E - E_r}{\frac{\Gamma}{2}}\right)^2}, \tag{1}$$

i.e., W(E) is halved for $E = E_r \pm \Gamma/2$, and, as shown in Fig. 1, decreases rapidly with increase in the difference $\Delta_r = E - E_r$. It is also easily seen that on both sides of resonance $W(E) \approx (\Gamma/2\Delta_r)^2$ for $\Delta_r \gg \Gamma/2$.

Equation (1) also gives the energy dependence of the cross sections for resonance absorption of light (or gamma quanta), which, from [41], are equal to

$$\sigma_a(E) = \sigma_0 W(E), \tag{2}$$

where

$$\sigma_0 = \frac{2I_e + 1}{2(2I_g + 1)} 4\pi \lambdabar_0^2, \tag{3}$$

where I_g and I_e are the total moments of the system in the ground and the excited state, and $2\pi\lambdabar_0$ is the wavelength at resonance. Because of the rapid drop-off from resonance on both sides of E_r, (2) and (3) neglect the change in $\lambdabar_0$ in the resonance range and simply use $\lambdabar_0$. Further, (2) is based on the assumption that the excited state may only be deactivated by re-emission of light to the ground state – in the general case there should be another factor Γ_γ/Γ (i.e., the ratio of the partial radiation width to the total width of the excited state) numerically equal to the probability of deactivation by reradiation of a light (or gamma) quantum. In nuclear gamma transactions with emission of gamma quanta, there is competition with internal electron conversion, in which the excitation energy is transferred directly from the nucleus to the electron shells and is used up in stripping off atomic electrons. The ratio of the probability of stripping off electrons to the probability of gamma emission is usually called the internal conversion coefficient α. It is obvious that the fraction of the decays of the excited state in which gamma quanta are emitted is given by $\Gamma_\gamma/\Gamma = 1/(1+\alpha)$, and, because of this fact, (2) must be rewritten in the form

$$\sigma_a(E) = \sigma_0 \frac{1}{1+\alpha} W(E). \tag{2'}$$

Integrating the absorption cross sections over the whole spectrum gives the quantity

$$\sigma_{int} = \int_0^\infty \sigma_a(E)\,dE = \frac{\pi}{2}\Gamma\sigma_0\frac{1}{1+\alpha} = \frac{\pi}{2}\Gamma_\gamma\sigma_0. \tag{4}$$

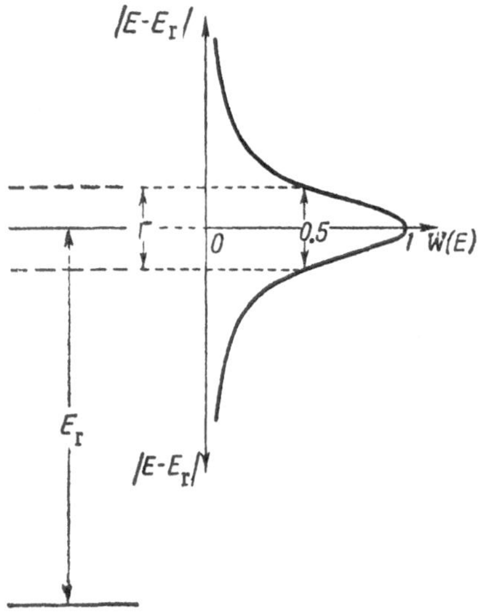

Fig. 1. A quantum transition and the natural width and shape of the resonance line.

The Role Played by Doppler Broadening and Recoil

It must be pointed out that, as a result of the so-called Doppler broadening of the resonance line the function $\sigma_a(E)$ observed experimentally is, in the general case, different from that given by (2). The atoms and molecules (or nuclei) emitting and absorbing the light (or gamma quanta) are not at rest, but are in thermal motion, and the light frequency ν for bodies moving at some velocity v and at an angle θ to the direction of motion of the quanta experiences a Doppler shift of the amount $\Delta\nu = -\nu(v/c)\cos\theta$ (where c is the velocity of light). Let the radiator emit strictly monochromatic quanta with energy E_r. To an absorber moving with the velocity v opposite to the quanta, the energy will seem to be $E_r(1 + v/c)$, while to the same absorber moving in the same direction as the quanta, it will seem to be $E_r(1 - v/c)$. It is obvious that, if the energy of the emitted gamma quanta now starts to vary while we are investigating the function $\sigma_a(E)$, even an infinitely narrow resonance line will be converted into a band with the base width $2(v/c)E_r$. Since, for thermal motion, $m\overline{v^2}/2 = kT$, where k is Boltzmann's constant, T is the temperature, and m is the mass of the light emitters and absorbers, the effective (Doppler) resonance width at half-maximum will be equal to

$$D = \frac{1}{c}\sqrt{\frac{2kT}{m}}\, E_r, \tag{5}$$

with, as a rule, $D \gg \Gamma$.

The Doppler resonance broadening naturally has no effect on the value of the absorption integral over the whole spectrum, which is equal, as before, to $(\pi/2)\Gamma_\gamma \sigma_0$. There is no displacement of the center of the resonance region in this case either. However, the maximum absorption cross section (at $E = E_r$) is now decreased by a factor of D/Γ.

We turn now to two typical examples from atomic and nuclear physics (see Table I): emission of the yellow D line of sodium and the gamma transition from the lower excited level of the Sn^{119} nucleus to the ground state (see Fig. 2a). Obviously, in both these cases the Doppler width is greater than the natural width. Since the Doppler width is proportional to the transition energy, the reduction in the maximum cross section in a nuclear transition is incomparably greater than in an atomic transition, and this in itself creates considerable difficulties in observing resonance gamma fluorescence. There is a second fundamental difficulty, however, which not only remains in force, but may even play the decisive role under hypothetical conditions (for example, at extremely low temperatures) where Doppler resonance broadening is completely absent. This difficulty is the loss of part of the energy in emission or absorption of a quantum to recoil of the emitter or absorber. By virtue of the law of conservation of momentum, the momenta of the emitter or the absorber will be equal to the momentum of the quantum, with the absorber momentum in the direction of motion of the quantum and the emitter momentum in the opposite direction. For a fixed emitter mass m, it is obvious that the following conditions will be satisfied:

$$E_r = h\nu + \frac{mv^2}{2}, \quad h\nu = mvc \tag{6}$$

(since $E_r \ll mc^2$, the rate of motion v of the nuclei or emitter atoms is always much less than c, and the nonrelativistic approximation may be used), from which

$$\frac{mv^2}{2} = R = h\nu\,\frac{h\nu}{2mc^2} \approx E_r\,\frac{E_r}{2mc^2}. \tag{7}$$

TABLE I

Transition characteristics	Atomic transition (sodium D line)	Nuclear transition (excitation of Sn^{119} nuclei)
Transition energy E_r, eV	2.1	23,800
Natural width of excited level Γ, eV	$4.4\cdot10^{-8}$	$2.4\cdot10^{-8}$
Resonance wavelength $2\pi\lambda\!\!\!^{-}_0$, cm	$5.89\cdot10^{-5}$	$5.3\cdot10^{-9}$
Maximum cross section* σ_0, cm^2	$1.1\cdot10^{-9}$	$8.8\cdot10^{-18}$
Doppler width D (at room temperature), eV	$3.3\cdot10^{-6}$	$1.6\cdot10^{-2}$
$\sigma_0(\Gamma/D)$, cm^2	$1.47\cdot10^{-11}$	$1.32\cdot10^{-25}$
Recoil energy per atom R, eV	10^{-10}	$2.5\cdot10^{-3}$
$\sigma_0(\Gamma/2R)$, cm^2	$1.1\cdot10^{-9}$†	$4.2\cdot10^{-23}$
Γ/E_r	$2.1\cdot10^{-8}$	10^{-12}

*The value of σ_0 is calculated from the formula

$$\sigma_0 = \frac{2I_e+1}{2I_g+1}\,\frac{2.45\cdot10^{-15}}{E_r^2(\text{keV})}\,\text{cm}^2.$$

†The real cross section is not greater than σ_0, since the formula $\sigma_0(\Gamma/2R)$ cannot be used for $\Gamma > 2R$.

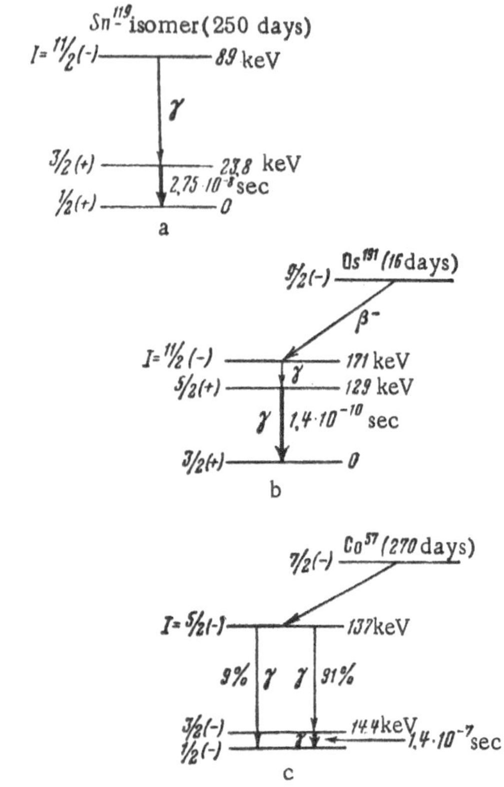

Fig. 2. Scheme of several nuclear transitions which exhibit the Mössbauer effect. The Mössbauer transitions are shown by heavy arrows, together with the transition time $\tau = \hbar/\Gamma$.

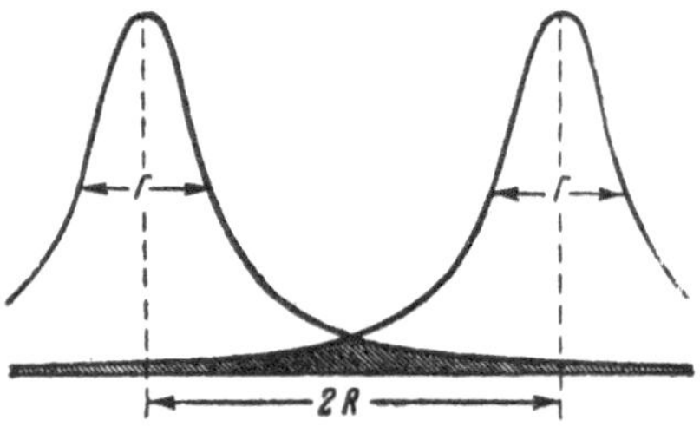

Fig. 3. Overlap of two resonance lines, moved apart as a result of recoil in emission and absorption.

Thus, the energy of the emitted quantum will be less than the resonance energy by an amount equal to the recoil energy $R \approx E_r^2/2mc^2$. On the other hand, to have resonance on absorption, the energy of the incident quantum must be (including loss of part of the energy to recoil of the absorber) greater than the resonance energy by the same amount R. As a result, the resonance maxima for the emitter and absorber are separated by the distance 2R, and if $2R \gg \Gamma$ resonance fluorescence cannot occur. Since the recoil energy R is proportional to the square of the resonance energy, R increases by an extremely large amount on going from the optical region of the spectrum to gamma radiation (Table I), and, while in the atomic case $R \ll \Gamma$, i.e., recoil has practically no effect on the resonance scattering and absorption cross section, for gamma fluorescence, we have instead $R \gg \Gamma$.

We can now rewrite (5) for the Doppler resonance width, in the form

$$D = 2\sqrt{RkT}, \tag{8}$$

from which it may be seen that the width is equal to twice the geometric mean of the recoil energy and the energy of thermal motion. Note further that if the emission and absorption lines are very far apart as a result of recoil ($R \gg \Gamma$), the role played by Doppler broadening is completely different from the case discussed above – the broadening no longer prevents but rather aids resonance fluorescence. As a matter of fact, as may be seen from Fig. 3, the cross section observed in resonance processes is determined in this case by the magnitude of the range of overlap of the two lines, of natural or Doppler width. The reduction in W(E) at the resonance maximum is accompanied by broadening of the lines and increase in the overlap region, which reaches its maximum value at $k(T_{em} + T_a) = 2R$.

The effective resonance fluorescence cross section, including recoil and Doppler broadening, is given by the formula

$$\bar{\sigma} = \frac{1}{2}\frac{\sigma_0}{1+\alpha}\frac{\Gamma}{2R}\left\{\frac{\Gamma}{2R} + \sqrt{\pi}\, g e^{-g^2}\right\}, \tag{9}$$

where

$$g^2 = \frac{R}{k(T_{em} + T_a)}.$$

It may be seen from Fig. 4 that even at the optimum temperature, $g = \frac{1}{\sqrt{2}}$, $\bar{\sigma} \approx 0.4\frac{\sigma_0}{1+\alpha}\frac{\Gamma}{2R}$, which reduces the gamma fluorescence cross section in our example of Sn^{119} by more than five orders of magnitude, while, in an atomic transition, recoil has no effect at all on the cross section for resonance processes.

The effect of Doppler broadening and the effect of recoil are also illustrated in Fig. 5, which, for emitters or absorbers with atomic weight A = 100 and a temperature of 300°K, gives values of R and D for various transition energies according to the formulas

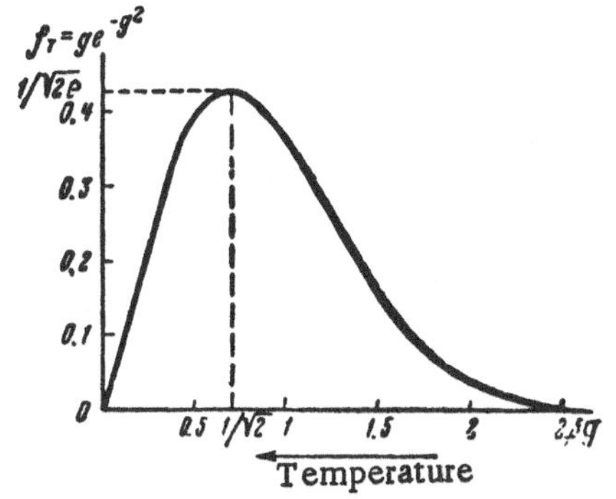

Fig. 4. The function ge^{-g^2}, giving the relation between the resonance fluorescence cross sections and the temperature resulting from Doppler broadening of resonance lines separated by recoil effects.

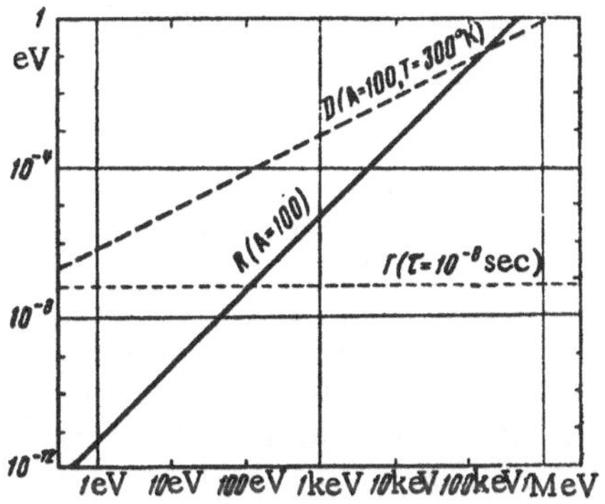

Fig. 5. Recoil energy and Doppler width as a function of transition energy. Atomic weight of emitter (or absorber) A = 100, temperature 300°K. The dotted line is the natural width of the level ($6.6 \cdot 10^{-8}$ eV), corresponding to the lifetime $\tau = 10^{-8}$ sec.

$$\frac{R}{E_r} = 5.37 \cdot 10^{-10} \frac{E_r}{A} \qquad (10)$$

and

$$\frac{D}{E_r} = 4.3 \cdot 10^{-7} \sqrt{\frac{T}{A}}, \qquad (11)$$

where E_r is expressed in electron volts. This example is taken from the review [26], where, however, both (9) [Eq. (63) in the review] and the line $D(E_r)$ (Fig. 1-*l* of the review) are given with some typographical errors that have been corrected here.

The decisive role played by recoil effects in the first unsuccessful attempts made to observe nuclear resonance fluorescence was first pointed out by I. Ya. Barit and M. I. Podgoretskii [42] in 1946. In the next few years, this fact was taken into account, and a number of papers appeared in which recoil was partially compensated for either by heating the source (for example for Hg^{198*}, where R = 0.46 eV and the optimum temperature is about 5200°C, the heating was carried to 1100°C), or by very rapid (up to $8 \cdot 10^4$ cm/sec) rotation of the source in the direction of the absorber, or, finally, by making use of the momentum acquired by the emitter nuclei in the radioactive decay act preceding gamma emission.

A detailed review of the "pre-Mössbauer" papers on nuclear resonance fluorescence was published comparatively recently by B. S. Dzhelepov [43]. Interest in these delicate and difficult experiments did not get beyond the walls of a few nuclear spectroscopic laboratories. Possibly the most important result of using nuclear resonance fluorescence with the recoil compensated for was the experiment of Goldhaber, Grodzins, and Sunyar [44], who found by using such compensation in the electron capture in Eu^{152} nuclei preceding emission of gamma quanta by Sm^{152} nuclei that the neutrino spin is oriented opposite to the direction of motion, i.e., that neutrinos are "left-hand screw" particles. However, even this experiment, in itself important and beautiful, did not open up any wide prospects for nuclear resonance fluorescence. The state of affairs did not change greatly until after Mössbauer's classical work [17-19].

CHAPTER II

THE MÖSSBAUER EFFECT

Mössbauer's Discovery

In 1957 the German physicist Rudolf Mössbauer began a study of resonance scattering of the 129-keV gamma quanta emitted by the excited Ir^{191} nuclei formed in beta decay of the mother isotope Os^{191} (see Fig. 2b). In this gamma radiation the recoil energy is R = 0.046 eV, so that the optimum emitter and absorber temperature for compensating for recoil is comparatively small and reaches a value of about 280°C. Even at room temperature the resonance lines of the emitter and absorber are quite strongly overlapped. Wishing to reduce the role of the overlap and measure the "background" of his apparatus in the absence of resonance absorption, Mössbauer placed both the emitter and the absorber in liquid air (at T ~88°K). He expected that the transmissivity of the absorber would increase [according to (8) the value of $\bar{\sigma}$ decreases by a factor of 20 when the temperature is changed from 300 to 88°K] and that the counting rate of the gamma counter placed behind the absorber would increase accordingly.

What happened was exactly the opposite – the counter gave fewer counts than without cooling, i.e., the gamma absorption increased.

Mössbauer very quickly grasped the significance of this result, which at first glance seemed paradoxical, since the theory of the newly discovered effect had already been worked out 20 years earlier in a paper by Lamb [45] on the capture of neutrons by atoms making up a crystal lattice. However, in all these 20 years, investigators working with neutrons and who were well acquainted with Lamb's work never suspected that it was possible to make a direct application of the effect to gamma fluorescence, while investigators working with resonance scattering and absorption of gamma quanta did not make use of results obtained in the neighboring field of nuclear physics – instructive example of the harm done by too narrow specialization!

Elements of the Theory of Recoilless Resonance Gamma Fluorescence

What, then, is the reason the gamma absorption increased when the temperature was lowered in Mössbauer's experiment [17]? The essence of the matter is that if an emitter or absorber atom forms part of a crystal lattice, there is no longer the unique relation between the momentum of the gamma quantum and the recoil energy which exists for free atoms. Since the recoil energy is far from being sufficient to rupture the chemical bonds in the lattice, it becomes the property of the lattice as a whole and can only be radiated in the form of collective excitation quanta or phonons. It is also possible that no phonons are excited, and then the recoil momentum, equal to the momentum of the gamma quantum, is taken by the whole lattice. But for this case, (5) no longer involves the mass of a single atom, but rather the mass of the crystal lattice; hence the actual magnitude of the recoil energy (which we shall call R_{cryst}, crystal recoil) – in contrast to the recoil energy of a single atom R – is infinitesimally small, much less than the natural level width ($R_{cryst} \ll \Gamma$); the recoil then ceases to prevent the observation of resonance scattering. Moreover, at the same time the recoil disappears, the Doppler broadening of the resonance line drops out. Both the recoil energy and the Doppler width are determined by the masses of the emitter and the absorber, and for this reason the Doppler width may also be simply expressed in terms of the recoil energy [cf. (8)]. Going from the recoil energy of a single nucleus R to the recoil energy of a whole crystal R_{cryst}, which satisfies both the inequality $R_{cryst} \ll \Gamma$ and the more rigorous condition $R_{cryst} \ll \Gamma(\Gamma/kT)$, means going to conditions such that the Doppler level width is much less than the natural width, i.e., an undisplaced resonance line occurs with its natural width. The probability of observing resonance gamma fluorescence on an undisplaced line (i.e., without recoil) depends on the normal vibration spectrum of the lattice and on the probability of exciting different levels of the vibrations.

The most widely used and simple way of describing the lattice vibration spectrum is by the so-called Debye approximation, in which the lattice has a continuum of oscillator states with different characteristic frequencies ω, up to some maximum frequency ω_{max}, which is determined by the number of oscillators N in the lattice volume V and characterized by the so-called Debye temperature θ:

$$\hbar\omega_{max} = k\theta. \tag{12}$$

The distribution function of the oscillator frequencies is assumed to be proportional to ω^2, while the normalizing condition is given in the form

$$\int_0^{\omega_{max}} 4\pi \frac{V\omega^2 \, d\omega}{(2\pi)^3} \frac{3}{u^3} = 3N, \tag{13}$$

where u is the mean velocity of sound in the solid in question (averaged over all directions of propagation and polarization states, with $3/u^3 = 1/u_l^3 + 2/u_t^3$, where l stands for longitudinal and t for transverse waves), and 3N is the total number of degrees of freedom in the lattice. Obviously, this gives

$$\omega_{max} = \left[6\pi^2 u^3 \frac{N}{V}\right]^{1/3} \tag{14}$$

or

$$k\theta = \hbar \left[6\pi^2 u^3 \frac{N}{V}\right]^{1/3}. \tag{15}$$

The function giving the distribution of characteristic frequencies may be expressed in the form

$$f(\omega)\, d\omega = 9N \left(\frac{\hbar}{k\theta}\right)^3 \omega^2 \, d\omega \quad \text{for} \quad \hbar\omega \leqslant k\theta \tag{16}$$

and

$$f(\omega)\, d\omega = 0 \quad \text{for} \quad \hbar\omega > k\theta. \tag{17}$$

In the simplest case of an isotropic cubic crystal, the maximum frequency ω_{max} and the corresponding Debye temperature are given approximately by the equation

$$\hbar\omega_{max} = k\theta \approx 2\pi \frac{u}{2d} \hbar, \tag{18}$$

where d is the lattice constant.

At extremely low temperatures (T = 0), the lattice executes only zero vibrations with some characteristic frequency spectrum. Obviously, in this case there can be no transitions of oscillators from higher vibrational levels to lower, since all the oscillators are thus found on the lower levels, while the upper levels are empty. Accordingly, there is no phonon absorption spectrum under these conditions, i.e., there is no energy transferred from the crystal lattice to a gamma quantum incident on it. But even at zero temperature there is still the possibility of oscillators jumping to higher levels, using for this purpose part of the energy of an incident gamma quantum, i.e., there is a definite phonon emission spectrum.

If the phonon excitation probability is large, the Mössbauer effect may for this reason be very weakly pronounced at low temperatures. However, no matter how things stand at low temperatures, the probability of recoilless resonance gamma fluorescence will drop with increase in temperature. Since the phonons obey Bose statistics (they are Bose quasi-particles), the greater the number of phonons already excited the higher will be the probability of exciting new phonons from the recoil energy in emission and absorption of a gamma quantum

In this case, the picture is completely analogous to the phenomenon of induced resonance emission of quanta [25].

After applying the Debye theory of a solid to the energy dependence of neutron absorption by atoms in crystals in the vicinity of a resonance level, Lamb showed that at low temperatures ($T \ll \theta$) an undisplaced resonance line will be observed which has the Doppler width $D = 2\sqrt{R\varepsilon}$ (instead of the energy of thermal motion kT, we have the mean energy ε of one vibrational degree of freedom of the crystal lattice). A similar consequence of the collectivization of the recoil energy in emission and absorption of gamma quanta is the intensification of the undisplaced line with its natural width when the emitter and absorber are cooled, which is the reason an increase in resonance absorption with a drop in temperature was observed in Mössbauer's experiment. A theoretical treatment of the probability f of a transition (in which an individual nucleus receives the recoil energy R) not accompanied by phonon excitation, i.e., by change in the internal energy state of the crystal lattice, is relatively simple to make in the Debye approximation and is characterized by the so-called Debye–Waller temperature factor:

$$f = \exp\{-2W(T)\}, \tag{19}$$

where

$$W(T) = \frac{3R}{k\theta}\left[\frac{1}{4} + \left(\frac{T}{\theta}\right)^2 \int_0^{\theta/T} \frac{x}{e^x - 1}\,dx\right]. \tag{20}$$

In the low-temperature range ($T \ll \theta$),

$$W = \frac{3}{4}\,\frac{R}{k\theta}, \tag{21}$$

while, for $R \lesssim 2k\theta$, f reaches values that are nearly unity, and for high temperatures ($T \gg \theta$)

$$W = \frac{3}{4}\,\frac{R}{k\theta}\,4\,\frac{T}{\theta}, \tag{22}$$

i.e., $W \gg 1$ and the factor f is accordingly very small.

The effective cross section at maximum of the Mössbauer resonance absorption line is given by the expression

$$\sigma_{max} = \sigma_0 f'(T')\,\frac{1}{1+\alpha}, \tag{23}$$

where $f'(T')$ is the Debye–Waller factor for an absorber with the temperature T'.

The classical theory of the Mössbauer effect developed by F L. Shapiro [22] leads to a very simple interpretation of the Debye–Waller factor in this case, namely:

$$f = \exp\left\{-\frac{\overline{x^2}}{\lambda\!\!\!^{-2}}\right\}, \tag{24}$$

where $2\pi\lambda\!\!\!^-$ is the wavelength of the resonance gamma quanta and $\overline{x^2}$ is the mean square deviation of a vibrating atom in the lattice from its equilibrium position (in the direction of observation).

Thus, the Mössbauer line is well defined if the amplitude of the vibration of the atoms in the lattice is small in comparison with the wavelength of the gamma radiation, which occurs principally at low temperatures.

Thus, at quite low temperatures, the spectra of the emission and absorption of gamma quanta by nuclei making up a crystal lattice consists of two components: a "trivial" broad component, due to thermal motion of the atoms in the crystal lattice, and to change in the vibrational states of the lattice during emission and

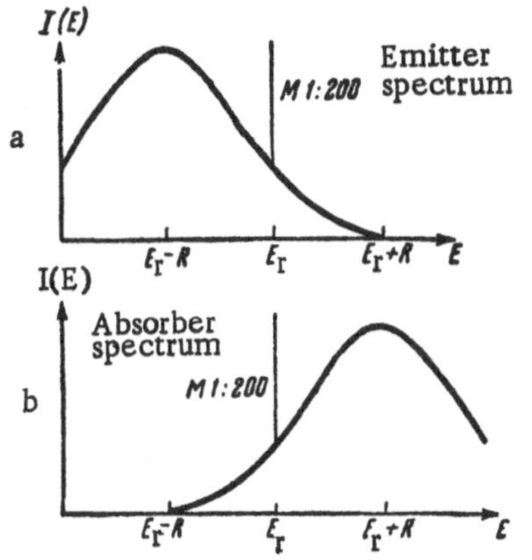

Fig. 6. Form of the emission and absorption spectrum of 129-keV gamma quanta for a transition in Ir^{191} nuclei at T = 88°K. The length of the line giving the recoilless resonance fluorescence contribution is reduced by a factor of 200 in the figure.

absorption of gamma quanta; a narrow resonance line of the natural width, due to recoil momentum taken up by the lattice as a whole, i.e., practically no energy is lost to recoil. The form of the two components of the emission and absorption spectrum for the 129-keV gamma transition in Ir^{191} at 88°K is shown schematically in Fig. 6. Increasing the temperature produces some increase in overlap of the broad components of the spectrum, which, however, is more than compensated for by a sharp reduction in the contribution from the undisplaced lines, with the result that the resonance absorption is not increased, but rather weakened.

Fundamentals of the Experimental Method of Observing the Mössbauer Effect

In subsequent simple and instructive experiments, Mössbauer [18, 19] finally showed that he had correctly interpreted as recoilless resonance gamma fluorescence the effect that he had observed; at the same time he laid the foundation for the experimental method used in all subsequent investigations of the phenomenon.

Since the ratio of the natural resonance width to the nuclear transition energy is exceedingly small (for Sn^{119}, for example, $\Gamma/E_r = 10^{-12}$ – see Table I), exceedingly small changes in the energy of the emitted gamma quanta are sufficient to disturb the resonance conditions substantially. In order to get the exceedingly small changes required, Mössbauer made use of the Doppler effect, but no longer at large velocities (up to hundreds of meters per second – of the order $(R/E_r)c$, as were used to increase the resonance fluorescence in the old experiment [43]) but at much lower velocities – of the order $(\Gamma/E_r)c$.

Schematic diagrams of such an experiment using the Doppler effect to vary the energy of the gamma quanta are shown in Figs. 7 and 8.

A gamma counter (for example, a scintillating crystal with a photomultiplier) detects the gamma quanta that are emitted by the source and have passed through the absorber; the latter may be moved relative to the source at a variable rate v (+v is opposite to and –v is in the same direction as the motion of the gamma quanta). Since, as the absorber is moved the frequency ν and the energy $E = h\nu$ of the incident gamma quanta vary on account of the Doppler effect,

$$\frac{\Delta E}{E} = \pm \frac{v}{c},$$

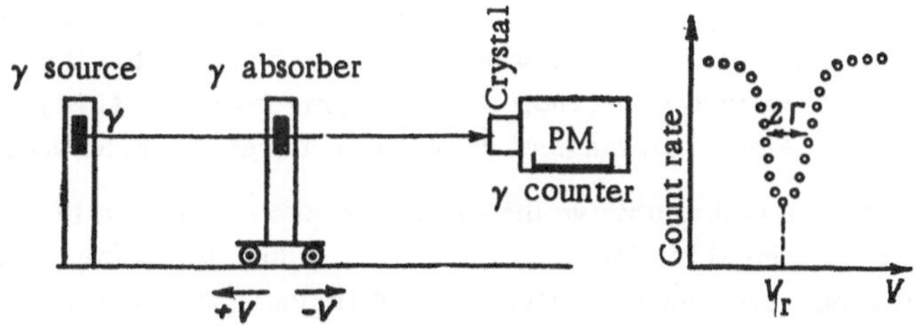

Fig. 7. Schematic diagram of an experiment on the Mössbauer absorption spectrum.

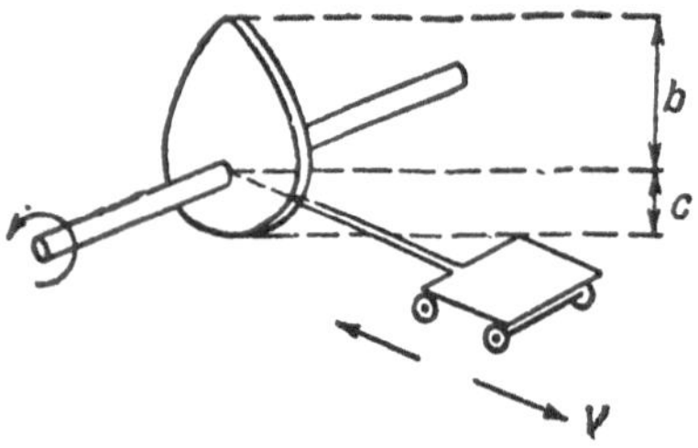

Fig. 8. Simplest scheme for transforming rotational to translational motion, giving a constant rate of motion in experiments on the observation of Mössbauer spectra.

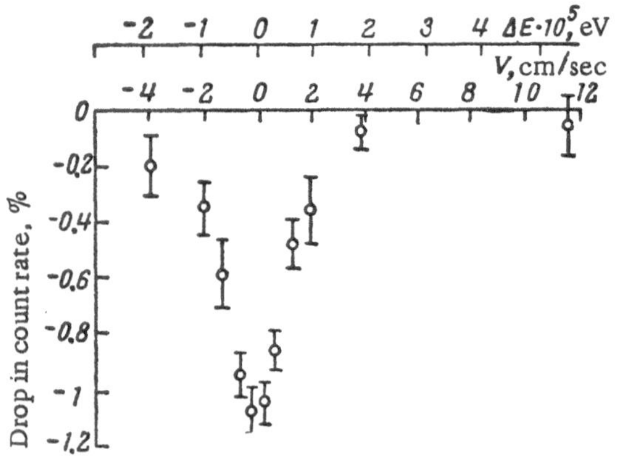

Fig. 9. Mössbauer absorption spectrum for gamma quanta for excited Ir^{191} in metallic irridium.

there is also a change in the conditions under which absorption occurs, as long as ΔE is comparable to Γ. As a result, the gamma count rate depends on the rate of motion of the absorber, as shown schematically (for thin source and thin absorber) in Fig. 7, and reaches its minimum value (maximum absorption) at the rate of motion corresponding to the center of the resonance line. If the emitter and absorber are identical, the center of the resonance line corresponds to $v = 0$ (absorber at rest), while in other cases the line is found to be displaced by the amount $\Delta E = E(v/c)$. If electric and magnetic fields are applied, the resonance line may be split, and the absorption spectrum then shows several resonance minima. A curve showing the counting rate of gamma quanta that have passed through the absorber or that have been scattered by it as a function of the rate of motion of the absorber with respect to the emitter may be given the name "Mössbauer spectrum." The first such absorption spectrum, obtained by Mössbauer himself [18, 19] in experiments with Ir^{191}, is shown in Fig. 9. A velocity of 1 cm/sec in these experiments corresponded to a change in gamma energy of $4.3 \cdot 10^{-6}$ eV. The shape of the absorption line made possible a direct determination of the natural resonance width ($\Gamma = 4.6 \cdot 10^{-6}$ eV) and, hence, of the lifetime of the excited state of Ir^{191} ($\tau = 1.4 \cdot 10^{-10}$ sec).

Thus, experiments for obtaining Mössbauer spectra simply involve finding the gamma absorption (or, less frequently, scattering) curve for the sample under test as a function of the rate of motion of the sample with respect to the source.

Without going into the details of construction of the various experimental setups, let us note simply that all these setups can be divided into two basic classes — those having a constant, and those having a variable, rate of motion. When a constant rate of motion is used, each experiment has a definite fixed rate of motion v of the absorber (less frequently of the emitter), so that half the time the absorber is moving toward the source (+v), and the other half it is moving away from the source (−v), and, at the instant the direction of motion changes, a measurement is made of the gamma count rate at zero velocity. Thus, each such experiment gives three points (+v, −v, 0) on the Mössbauer spectrum. When a variable velocity is used, in each period of time T the velocity rises smoothly from zero to some maximum positive value $+v_{max}$ ($t = T/4$), again drops to zero ($t = T/2$), becomes negative, reaching the value $-v_{max}$ ($t = 3T/4$), and, finally, again returns to zero ($t = T$). Accordingly, a single experiment gives the whole Mössbauer spectrum curve at once. However, to do this means using special electronic devices, namely, multichannel analyzers. Depending on the number n of channels in the analyzer, each period T of velocity change is divided into n intervals, so that the i-th channel of the analyzer records the number of counts received in the short period of time from $t_b = (i-1)T/n$ to $t_f = (i/n)T$ of each period, which thus corresponds to some small range of velocities between $v(t_b)$ and $v(t_f)$.

The constant rates of motion used to observe the Mössbauer spectra are usually obtained by the use of various devices that transform rotational motion to linear motion. Thus, for example, the cam ($b - c = 40$ mm) shown in Fig. 8, rotating at n rpm gives the cam follower a rectilinear motion of ±1.33n mm/sec. Variable rates of motion may be obtained either by rotating cams (but of a different shape) or by a source of sinusoidal vibrations, for example, by transforming the vibrations of a dynamic loud speaker into translational motion of an absorber connected to it.

Depending on the sharpness of resonance, i.e., on the value of the ratio Γ/E_r, the velocity used in working with different isotopes may vary from centimeters per second down to exceedingly small and difficultly controllable values of the order of fractions of a micron per second:

$$\text{for } Zn^{67} \quad \frac{\Gamma}{E_r} = 5.2 \cdot 10^{-16}, \text{ i.e., } \frac{\Gamma}{E_r} c = 0.15 \quad \mu/\text{sec.}$$

Illustrations of the General Importance of the Mössbauer Effect

Before we go to actual chemical applications of the Mössbauer effect, let us consider — although only briefly and by no means exhaustively — some examples which illustrate the general importance of the effect in a number of branches of physics and engineering, as well as the importance of the results that it has already yielded and the new possibilities that are opening up. The example of Ir^{191} has already been used above as demonstration of a simple determination of the lifetime of very short-lived excited states of atomic nuclei from the shape of the Mössbauer spectrum. In addition to the lifetime, the Mössbauer effect makes it possible to find the decay scheme. This is due to the fact that resonance fluorescence can only be observed for a given "sort" of gamma quanta, namely if the quanta are being emitted in a transition to the ground state, but not to some intermediate excited state of the atomic nucleus, since it is only then that the quanta will be in resonance with absorber nuclei that are in the ground state. In both of the examples cited, although the Mössbauer effect provides a very convenient method, it is certainly not the only possible method of experimental study. There are, however, a number of completely unique possibilities which could not even have been thought of in nuclear physics before Mössbauer's discovery. We are speaking here of direct observation of nuclear level splitting in external electric and magnetic fields. While it was possible previously to determine the energy of gamma quanta in the best case with an accuracy of the tenth of a percent, the discovery of the Mössbauer effect has raised the accuracy by a factor of a billion! As a result, it has become possible to make a completely clear quantitative study of Zeeman splitting and, by means of this, to determine the magnetic moments of nuclei in excited states (μ_e). As a matter of fact, even if the magnitude of the magnetic field H producing the Zeeman splitting is not known in advance, it may easily be found first by analyzing the effect of the nuclear term splitting in the ground state, for which the spin I_0 and the magnetic moment μ_0 are known, on the Mössbauer spectrum since

$$H = \frac{\Delta_m^0 I_0}{\mu_0}, \qquad (25)$$

where Δ_m^0 gives the equidistant spacings between the $(2I_0 + 1)$ components of the Zeeman term splitting of the ground state of the nucleus. Having found H in this way, and using the multiplet order $(2I_e + 1)$ of the excited state term to calculate the spin I_e, we can easily find the magnetic moment μ_e from (25), which is now applied to the excited nucleus. Possibilities are also opened up for determining the polarization of the gamma radiation for each of the Zeeman splitting components.

Similar considerations apply to the quadrupole splitting of nuclear levels in nonuniform electric fields and to the determination of nuclear quadrupole moments, about which more will be said below.

The matter of the chemical shift of Mössbauer spectrum lines will also be discussed in detail in the rest of the book. However, since we are considering the importance of the Mössbauer effect in nuclear physics, we should mention here the possibility of using the chemical shift to find the difference between the nuclear radii in the excited (R_e) and in the ground (R_0) state.

The prospects for using the Mössbauer effect in solid state physics are very attractive. Here we should mention measuring the local magnetic and nonuniform electric fields as a function of the structure and composition of the lattice and of various external conditions (such as temperature and pressure), in which measurements of the polarization of the various Zeeman components may be used to find the direction of the internal magnetic fields in the domains of ferro- and antiferromagnetics. We can mention further investigation of other defects, and of Rayleigh scattering of gamma quanta in solids. The Mössbauer effect, as will be shown later on, makes it possible to find the contribution of the optical branches to the vibration spectra of lattices, and it is not impossible to make a direct determination of the form of the phonon spectra by means of resonance gamma fluorescence. A number of interesting proposals include using the Mössbauer effect to measure ultra-low temperatures (of the order of 10^{-3}°K) and to investigate superconducting states. The use of the Mössbauer effect in solid state physics has been discussed in special detail in the view by Yu. M. Kagan [25], who has made a considerable contribution to the field.

In passing to the importance of the Mössbauer effect for general physics, we must first of all stop to consider the classical experiment of Pound and Rebka [23, 46] on finding the weight of photons. It follows from Einstein's equivalence principle that the energy of gamma quanta, like that of ordinary ponderable bodies, will depend on the gravitational potential of the earth; hence, if they are raised to a height h, the frequency of gamma quanta will decrease by the amount $\Delta\nu = \nu(gh/c^2)$, which amounts to about $10^{-16}\,\nu(l)$ per meter of height. This difference is so infinitesimally small that no one had any hope of observing it until many years hence by comparing the time shown by very accurate atomic clocks on satellites with that shown on the earth. However, thanks to the Mössbauer effect, by using the source Fe^{57} ($\Gamma/E_r = 3 \cdot 10^{-13}$) and comparing the absorption under conditions in which either the source or the absorber was on the top of a tower 21 m high, Pound and Rebka were able both to observe the resulting gravitational resonance shift and to show that the shift is in quantitative agreement with Einstein's theory. This brilliant experiment, outshines even the most delicate experiments, which show, for example, that it is possible to use the Mössbauer effect to observe the second-order Doppler effect, as well as small accelerations in the motion of the source with respect to the absorber, or which make it possible to convince oneself of the rigorous equivalence of inertial and gravitational mass, i.e., with no anisotropy present in the inertia, which could occur as a result of the asymmetry in the distribution of matter with respect to the earth.

In order to become acquainted in more detail with the outlook for various physical applications of the Mössbauer effect, we refer the reader to other reviews, for example, to the excellent and quite popular and well-illustrated review by Frauenfelder [26], and we shall recount here merely some of the most general possibilities for using the effect in engineering: defectoscopy of all possible types of materials, which shows not merely the presence of flaws, but the presence of internal stresses in the materials; detection of exceedingly weak vibrations, of displacements, and of changes in velocity or acceleration; developing portable equipment for radio-activation analysis, which is completely specific in its sensitivity and therefore not bothered by background, for the purpose of detecting the presence of Mössbauer elements (for example, iron, tin, gold) in various ores and minerals, as well as for assaying such elements.

CHAPTER III

SHIFT AND SPLITTING OF MÖSSBAUER SPECTRUM LINES

The analog of the isotope, or, better, optical spectrum isomer shift in Mössbauer spectra is the shift which is also usually called isomeric, but which is more correctly called chemical in the present case.

The minimum total energy of a system made up of a nucleus and electron shells occurs for the case of a point nucleus. The farther the electric charge distribution of the nucleus extends, the higher – for a given electron shell structure – the total energy of the system. Accordingly, an increase in the dimensions of an isomeric excited state of a nucleus over the dimensions in the ground state produces an increase in the gamma transition energy. On the other hand, the greater the electron density $|\psi(0)|^2$ at the nucleus, due principally to s-electrons (the contribution of $p_{1/2}$ electrons is considerably less), the greater also – for given nuclear dimensions – the total energy of the system. Accordingly, changing $|\psi(0)|^2$ in the atoms of the absorber from what it is in the emitter also causes them to have different gamma transition energies (see Fig. 10).

The change in transition energy resulting from a change in nuclear dimensions between the ground and the excited state is given by the relation [47]

$$\Delta E = \pi e^2 \frac{a_0^2}{Z} |\psi(0)|^2 R^{2\rho} \frac{\Delta R}{R} \left(\frac{2Z}{a_0}\right)^{2\rho} \frac{1}{\Gamma(2\rho)^2} \frac{3-2\rho}{3+2\rho}, \qquad (26)$$

where a_0 is the Bohr radius, Z is the nuclear charge, $\rho = \sqrt{1-(Ze^2/\hbar c)^2}$, R is the mean radius of the nuclear electric charge distribution, and $\Delta R = R_e - R_0$ is the change in this radius on going from the ground (R_0) to the excited (R_e) state.

The experimentally measured chemical shift, $\delta = \Delta E_a - \Delta E_{em}$ is proportional to

$$\delta \sim (R_e - R_0)\{|\psi(0)|_a^2 - |\psi(0)|_{em}^2\}, \qquad (27)$$

where a positive shift means that, at the absorption maximum in the experiments in which the spectrum is being measured, the absorber is moving toward the emitter. In addition to the chemical shift, an additional shift is, generally speaking, possible in the Mössbauer lines, due to the so-called second-order Doppler effect [48, 49]. If we write the Doppler frequency change in the gamma quanta in general relativistic form, including the time contraction for moving bodies, the additional term $1/2(v/c)^2$ is added to the right-hand side of the usual expression $\Delta\nu/\nu = -(v/c)\cos\theta$. For a periodic vibration in a lattice of Mössbauer atoms with velocity v(t), the usual Doppler term [v(t) c] cos θ is averaged and is equal to zero (if the period of the vibrations is much less than the lifetime of the excited state), but the second-order Doppler effect persists and shifts the Mössbauer line by the amount

$$\delta_{Dopp} = \frac{1}{2}\frac{\overline{v^2}}{c^2} E_r = \frac{\overline{E}_{kin}}{mc^2} E_r, \qquad (28)$$

where $\overline{E}_{kin}$ is the mean kinetic energy of the Mössbauer atoms in the lattice. If the emitter and absorber are in the same chemical states and at the same temperature, the Doppler shift in the gamma resonance energy is the same for both emitter and absorber, and no complications arise. But if there is a difference in chemical states

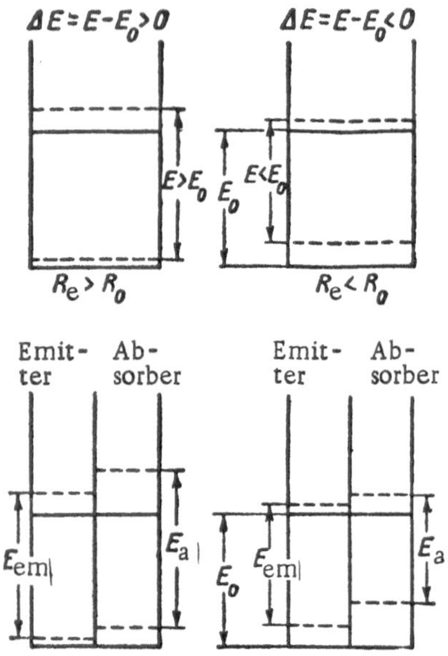

Fig. 10. Illustrating the changes in energy E of an electron–nucleus system when the dimensions of the nucleus and electron density in the vicinity of the nucleus change.

[in addition to the relation with $|\psi(0)|^2$ and because of differences in E_{kin}] of the emitter and absorber, and particularly in the temperatures, there is an additional Mössbauer line shift which, although it is much weaker than the chemical shift, is sometimes comparable to it. Accordingly, in precision measurements of Mössbauer spectra, account must be taken of the contribution of the second-order Doppler effect, which may be evaluated by, for example, varying the temperature difference between the emitter and the absorber.

It may be seen from (24) and (28) that the probability of the Mössbauer effect is given by the mean square displacement, while the Doppler temperature shift is given by the mean square rate of motion of a vibrating atom with respect to its equilibrium position. Accordingly, combining measurements of f and δ_{Dopp} opens up very wide, but so far unused, prospects for investigating the properties of crystal lattices.

We return now to the actual chemical shift.

Using (27) makes it possible first of all to find the sign of the quantity ΔR by comparing the chemical shifts of several absorbers for which it can be said from general chemical considerations which molecules have higher and which have lower electron density at the nucleus. Thus, the very first case of chemical shift, observed by Kistner and Sunyar [50] – the positive shift of Fe_2O_3 with respect to metallic iron-bearing in mind that the oxidation of Fe to Fe_2O_3 is accompanied by removal of the external (4s) electrons, which means that there is a reduction in $|\psi(0)|^2$, led these authors to the conclusion that ΔR was negative (i.e., $R_e < R_0$) for Fe^{57}.

In an entirely similar way, the positive chemical shift of bivalent tin compounds with respect to tetravalent tin derivatives, in which $|\psi(0)|^2$ should be less, led V. S. Shpinel' and co-workers [51] to the conclusion that $\Delta R > 0$ (i.e., $R_e > R_0$) for Sn^{119}, which was later confirmed quantitatively by comparing the shifts for quadrivalent tin compounds of the type SnR_4 with a different but unknown amount of ionic character in the Sn–R bond [52].

Especially great possibilities are opened up by (27) in conjunction with calculating the absolute values of $|\psi(0)|^2$. Thus, for example, the calculations of $|\psi(0)|^2$ for states of iron with a different number of 3d-electrons made by Watson and Freeman [53, 54] by the Hartree–Fock method formed the basis for determining the absolute value of ΔR/R for Fe^{57} and for the subsequent calculations of the contribution made by 4s-electrons to the valence bonds of iron in its various compounds [55].*

Later, we shall give a more detailed discussion of both these calculations and similar considerations related to our observations of the Mössbauer effect in tin compounds [52].

A second source of information on the electron shell structure of Mössbauer atoms is provided by the quadrupole splitting of the spectrum lines.

The interaction between a nucleus with the quadrupole moment $Q[cm^2]$ and an axially symmetric inhomogeneous electric field with gradient $q[V/cm^2]$ splits the level with moment I into sublevels with different magnetic quantum numbers m displaced by

$$\Delta E_I = eQq\,\frac{3m^2 - I(I+1)}{4I(2I-1)}\ \text{eV} \tag{29}$$

*It should however be kept in mind that there is no a priori reason the value of $|\psi(0)|^2$ obtained by Watson and Freeman, which is unquestionably the best for calculations on atoms, should be equally good for calculations on molecules.

(where e is expressed in electron charge units). In the most common case of Mössbauer M1 gamma transitions between $I = 1/2$ and $I = 3/2$, splitting the level with $I = 3/2$ into sublevels with $m = \pm 3/2$ and $m = \pm 1/2$ produces a spectral doublet in which one of the lines corresponds to the σ transition ($\pm 1/2 \rightarrow \pm 1/2$), while the other line corresponds to the π transition ($\pm 1/2 \rightarrow \pm 3/2$), and the difference between the energies of the π and σ transitions is $\Delta = 1/2\ eQq = 1/2 W$.

Here, the energy of the π transition is greater than the energy of the σ transition if W is positive, and less if W is negative. The magnitude of the quadrupole interaction constant W is itself only a relative measure of the inhomogeneity of the electric field at the Mössbauer nucleus. In order to compare this constant with the values of the field gradient with different assumed concrete electron shell structures, we have to make sometimes quite complicated calculations of the intramolecular electric fields (for example, in the sense of Townes and Dailey's calculations [56]) and find the nuclear quadrupole moment (esperimentally or also by calculation) for the nucleus in an excited state, as, for example, for Sn^{119} and Fe^{57}. So far, the analysis of the data on the quadrupole splitting of Mössbauer spectra has only been of a semiquantitative nature. However, a number of interesting results have been obtained for several iron and tin compounds. Let us illustrate briefly what we have just said by the case of tetravalent tin compounds. In symmetric compounds of the type SnX_4, where all four tin bonds are equivalent, the molecule is spherically symmetrical, and there is, as a rule, no quadrupole splitting. Here, the only measure of the degree of ionic character of the Sn–X bonds when investigating Mössbauer spectra is the chemical shift – the stronger the ionic character of the bond (in which Sn appears as an electron donor and X as an acceptor), the more the electrons are drawn away from the tin atoms and the lower the value of $|\psi(0)|^2$. If, however, a study is being made of an unsymmetric compound of the type SnR_iX_{4-i}, the Sn–R and Sn–X bonds are, in the general case, ionic to a different extent; hence even for complete $5s5p^3$ hybridization the molecule is no longer spherically symmetric, and the Mössbauer spectrum shows quadrupole splitting (for example, [52]). Any change in structure of the molecule (for example, replacing R by R' without changing X, or replacing X by X' without changing R) increases or decreases the ionic character of the tin bonds, which, in the present case, changes both $|\psi(0)|^2$ (i.e., the chemical shift) and the electric field gradient in the region of the tin nucleus (i.e., the quadrupole splitting).

The quadrupole splitting is often even more sensitive to the nature of the chemical bond (for example, the ionic character of the bond) than the chemical shifts of the Mössbauer spectrum lines. Further, the data on the chemical shifts and the quadrupole splitting sometimes complement one another to a considerable extent. Thus, for example, in the spectrum of compounds of the type R_2SnX_2, where R plays the role of donor with respect to the tin while X is an electron acceptor, the quadrupole splitting will, in the general case, increase in going either to substituents R' which are more electropositive than R or to substituents X' that are more electronegative than X. However, in the first case, the density of the electron cloud in the vicinity of the tin nucleus is more likely to increase, while in the second case, it is more likely to decrease.

We shall return to this question again in Chapter VII, but first we must discuss the general reasons it is possible to apply the Mössbauer effect to chemistry, and we shall stop to consider some of the fundamental difficulties that arise and show how they may be overcome.

In concluding this section we shall say a few words on the magnetic splitting of Mössbauer lines.

The Zeeman splitting picture in Mössbauer spectra is often quite complicated in view of the fact that in the general case it is determined by two values of the nuclear magnetic moments – in the ground and in the excited state – and the two (different) values of the local magnetic fields for the nuclei of the emitter and absorber. In addition, the various Zeeman components with different m are shifted differently by the quadrupole interaction. Nevertheless, interpreting the Zeeman splitting is much simpler than in the case of quadrupole splitting and, if the necessary set of experiments has been performed (without special quantum mechanical calculations) reduces to a direct determination of the absolute values of the magnetic field in the region where the Mössbauer nuclei are located.

CHAPTER IV

GENERAL REASONS IT IS POSSIBLE TO APPLY THE MÖSSBAUER EFFECT IN CHEMISTRY

Frequency of Occurrence of Mössbauer Transitions

The first question of importance in evaluating the prospects for making chemical applications of the Mössbauer effect is that of the frequency of occurrence of Mössbauer nuclei. As we know, the probability of observing the effect rapidly (exponentially in the Debye approximation) decreases with increase in the recoil energy $R = E^2/2mc^2$, which is proportional to the square of the gamma-transition energy E and inversely proportional to the mass of the emitter. In addition, as a result of the general reduction in nuclear level density for a given excitation energy with decrease in nuclear mass, high-energy (of the order of MeV) gamma-transitions predominate for light nuclei. Accordingly, we must exclude from the number of even potential Mössbauer emitters and exciters such chemically interesting light-element nuclei as carbon, nitrogen, and oxygen, not to mention, of course, hydrogen. The Mössbauer transition energies lie in a range from several keV (with no lower limit in principle) up to several hundred keV. The duration of the corresponding transitions will be neither too small — where the ratio of the level width Γ to the transition energy E become so large that the resonance loses its selectivity — nor too large, for very small values of Γ/E give low probability and poor observability of recoilless gamma scattering and absorption. The most promising range of values of the ratio Γ/E is 10^{-10}-10^{-14}, although observations of the Mössbauer effect have already been made for $\Gamma/E = 5.2 \cdot 10^{-16}$ (Zn^{67}) [57, 58]. It is naturally undesirable in this case to have large values of the conversion coefficient α, since if conversion electrons are emitted instead of gamma quanta the possibility of observing resonance fluorescence of course drops off. Lists of possible Mössbauer nuclei are given in a number of reviews [21, 26], and are reproduced in the Appendix with the addition of several examples from the heavy and transuranium elements. Although the group is comparatively restricted (different isotopes of 35 elements are mentioned, from iron to mercury), quite ample possibilities are opened up for various chemical studies; of particular interest, from the point of view of the chemist, is the use of the Mössbauer effect in the study of complex and metallo-organic compounds. We shall now give some of the elements for which recoilless resonance gamma fluorescence has been observed (underlined) or should be observed: <u>iron</u>, <u>nickel</u>, <u>zinc</u>, germanium, ruthenium, <u>tin</u>, antimony, <u>tellurium</u>, <u>iodine</u>, cesium, almost all the lanthanides (the effect has already been observed for Sm, Dy, Er, Tm), hafnium, tantalum, tungsten, <u>rhenium</u>, osmium, <u>iridium</u>, platinum, <u>gold</u>, and mercury. To this list must be added almost all the transuranium elements.* To the advantages of the Mössbauer effect over the NQR (Nuclear Quadrupole Resonance) method must also be added a possibility of using quadrupole moments not only for ground states, but for excited nuclear states as well. For example, the nuclei of the stable isotopes of iron, tin, tellurium, and tungsten have no quadrupole moment at all in the ground state, so that the NQR method is completely inapplicable to these elements. However, in recoilless resonance fluorescence it is possible to have quadrupole splitting of the excited levels with $I \geq 1$ (or $I \geq 3/2$) in all these cases.

---

*In the case of the transuranium elements, the gamma energies in the Mössbauer transitions are so close to the x-ray energies of heavy atoms that strong absorption may occur, which leads to additional experimental difficulties. Further, observations of the Mössbauer effect for γ quanta emitted immediately behind α particles may be interfered with by local heating of the material in the track of the α particles.

Role Played by a Light-Atom Environment

The next important question is whether or not it is possible to observe the Mössbauer effect in compounds in which the Mössbauer atoms are embedded in molecules or lattices constructed mainly of much lighter atoms such as hydrogen, carbon, nitrogen, oxygen, etc. In view of the fact that recoil of the Mössbauer nucleus produces combined oscillation of a large number of atoms, it would seem that the probability of observing recoilless gamma resonance in complex systems would be determined in the Debye approximation not by the masses of the emitters and absorbers themselves but by those of the basic, lighter components, i.e., that the recoil energy R in the Debye–Waller factor [see (19)-(22)] is not equal to $E^2/2mc^2$, where m is the mass of the Mössbauer atom itself, but to $E^2/2\overline{m}c^2$, where $\overline{m}$ is the averaged mass of the atoms in the lattice. If the light atoms were not only incapable of exhibiting the Mössbauer effect themselves, but in addition annihilated the effect by their presence, the changes of using recoilless resonance fluorescence in chemistry would be poor indeed. However, it has also been pointed out in the review by F. L. Shapiro [22] that finding the probability of the Mössbauer effect from the value of the mass averaged over the masses of all the atoms in the lattice is based on neglecting the role played by the high-frequency part of the optical branches of the vibration spectrum, the justification for which is far from obvious. The high-frequency optical vibrations represent considerable displacements of the Mössbauer atom with respect to its nearest neighbors, and temperature excitation of the vibrations begins to show up at considerably higher temperatures than is the case for acoustic phonons, the frequency spectrum of which starts at zero.

A consistent theory of the role played by the optical branches of the phonon spectra in the Mössbauer effect has been worked out by Yu. M. Kagan [59, 60], who showed that if the contribution from these branches is large it is incorrect to use the Debye temperature as even an approximate characteristic of recoilless gamma fluorescence. As a result of the large amplitudes of the optical branches, there will be both a large probability of the Mössbauer effect in light systems with heavy emitters or absorbers and an anomalously slow temperature dependence of the effect. Actually, as was first shown in the experiments of Wertheim [61] and Ruby [62], there is even a considerable probability of observing the Mössbauer effect in Fe^{57} in ferrocyanides, for example in $Na_4Fe(CN)_6 \cdot 10H_2O$, where the mean mass of the atoms is 2/11 of that of the atoms absorbing the γ quanta. However, these experiments, which were only made at one temperature – and that a quite low temperature (78°K) – have still not made it possible to tell anything about the general properties of the effect when the Mössbauer nuclei are in a light environment. Accordingly, the results of the experiments made by V. V. Sklyarevskii, B. N. Samoilov, and E. P. Stepanov on Dy_2O_3 [63], and by V. S. Shpinel' and co-workers on SnO_2 [64], as well as the later data on Eu_2O_3 in [65] are of more interest. All these oxides showed both a large probability of the Mössbauer effect and an anomalously weak temperature dependence. Also instructive in this respect is the result of our work [66], in which it was shown that there is quite a large Mössbauer effect in the polymer containing tin (based on methylmethacrylate) of the composition

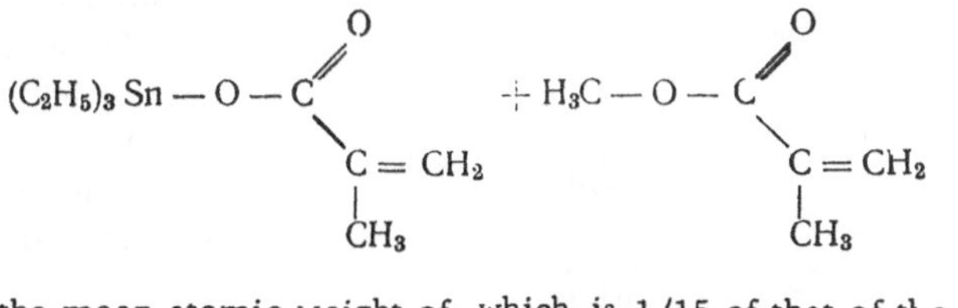

the mean atomic weight of which is 1/15 of that of the tin. The Mössbauer spectrum of this polymer at T = 77 and 195°K is shown in Fig. 11. The recoilless resonance absorption probability f' was reduced by not more than a half in this range of temperatures. The theoretical and experimental data just given are sufficient to eliminate

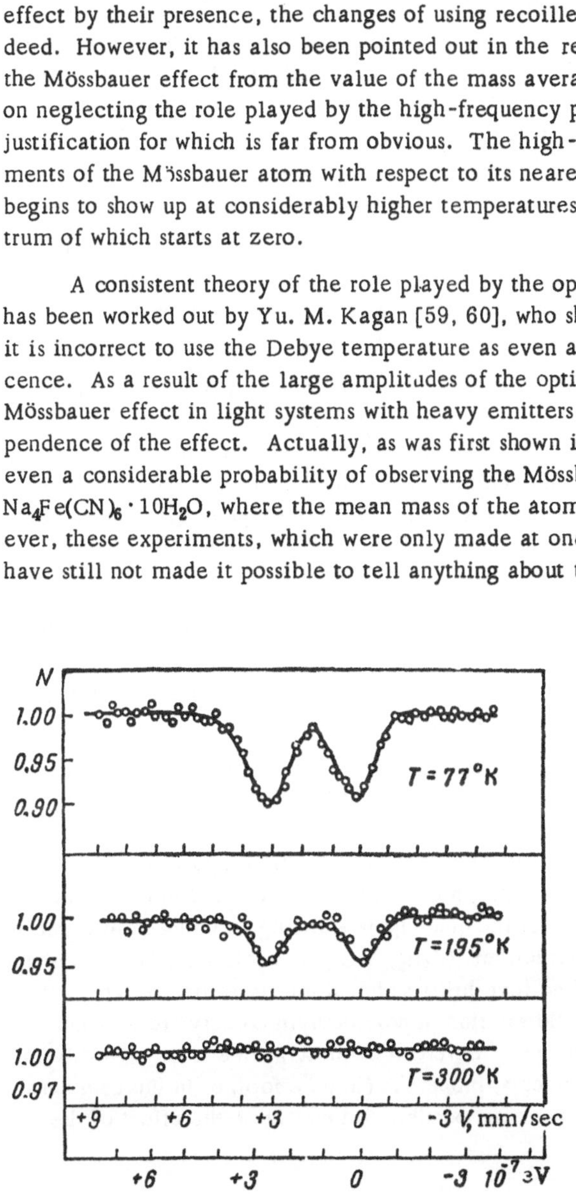

Fig. 11. Mössbauer spectra for a polymer containing tin (based on methylmethacrylate) at different temperatures. Source – SnO_2.

any skepticism with regard to the possibility of applying the Mössbauer effect to a very large group of chemical compounds.

Delimitation of the Roles Played by the Nearest Chemical Bonds and by the Macrostructure

One more question of importance for the outlook for chemical applications is that it is possible to separarate the features in the Mössbauer spectra that come from the structure of the various molecules under study, from those which come from the macroscopic properties of the lattice; the latter, of course, themselves depend to some extent on molecular structure. It has been demonstrated beyond doubt a number of times that in the general case some role is played by the macrostructure and that there is a definite effect of intermolecular and collective interactions, i.e., of the charge fields and magnetic moments occurring beyond the immediate vicinity of the Mössbauer atoms. It is sufficient to recall, for example, the differences in quadrupole splitting between tetragonal ($W = 2.5 \cdot 10^{-7}$ eV) and rhombic ($W = 3.5 \cdot 10^{-7}$ eV) tin oxide, SnO, or the difference (of $4.7 \cdot 10^{-8}$ eV) between the chemical shifts in white and gray tin [67]. The macrostructure effect shows up in the temperature changes in the magnitude of both the chemical shift and the quadrupole splitting. It is, however, also immediately obvious that "inert" systems may be found in which the macrostructure does not prevent a study of the properties of the individual molecules, which are of most interest to chemists.

As an illustration of these possibilities, we cite some results which we have obtained from observing the Mössbauer effect in tin compounds [68]. Figure 12 shows the curve found in [68] for a glass containing 9.1% SnO_2 along with silicon, boron, sodium, and aluminum oxides, the radiator also being crystalline tin dioxide, SnO_2.

In addition to the fact that there is a strong effect at 77°K in the absorber, which until recently was assumed to be typically amorphous, it is interesting to see that there is no chemical shift, in spite of the transition from pure SnO_2 to solid solution. Here the solid and involatile oxide SnO_2 does not exist in the form of separate molecules (of CO_2 type), but in the form of a single macromolecule of crystalline or polymer type, in which there is a network of $-Sn-O-$ bonds. It is clear from the results given that when the type of glass under study was being made there was no destruction of the continuous $-Sn-O-$ bonds, and the original structure of the SnO_2 inclusions was maintained.

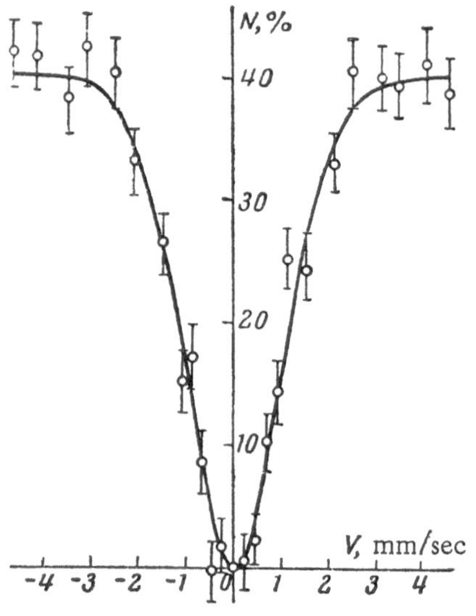

Fig. 12. Mössbauer spectrum for a glass containing tin (9.1% SnO_2) at 77°K. Source – SnO_2.

The purest example of "inert" matrices, in which the spectral characteristics are determined by the properties of the individual molecules and are only slightly distorted by macrostructure factors, is that of molecular crystals.

In this connection there is interest in the results of our experiments on organotin compounds [68] and, in particular, in comparing the Mössbauer spectra for molecular crystals of organotin compounds with those for their solutions and various inert organic solvents. Thus, no changes were observed in either the chemical shift or the quadrupole splitting on going from crystalline diethyltin dichloride* $(C_2H_5)_2SnCl_2$ to a solution of dichlorethane. The same chemical shift was found for crystalline tetraphenyltin and its solid solution in polymethylmethacrylate polymer. This is of double interest to chemists, since it is possible to investigate molecular properties (in pure form or in inert solvents) and at the same time possible to investigate the effect of the

* The names of the organotin compounds are given in accordance with the terminology used in the book "Organo-Tin and Organo-Germanium Compounds" by R. Ingham, S. Rosenberg, H. Gilman, and F. Rikens (Russian translation edited by I. F. Lutsenko, Moscow, Foreign Literature Press, 1962).

solvents themselves as a function of polarity, their ability to enter into a chemical reaction with the dissolved substance, form complex compounds with it, etc.

There are a number of cases in which it is possible to eliminate the effect of the macrostructure in the Mössbauer spectra and, at the same time, detect the role played by the chemical bonds nearest the absorber (or emitter) nucleus. This is possible when the distant atoms or functional groups are practically without effect on the ionic character of the valence bonds of the Mössbauer atom itself, i.e., they change neither the chemical shifts nor the quadrupole splitting of the spectrum lines in the manner discussed in Chapter III. In other words, separating out the effect of the bonds nearest the Mössbauer atoms "in pure form" is more probable in those cases where the distant atoms or groups do not exert a strong inductive or mesomeric rearrangement effect on the electron shells, and the Mössbauer atom itself or the functional group in which it is located has very weak inductomeric polarizability.

An example of separating out the effect of the nearest chemical bonds in this way is provided by the fact that practically identical Mössbauer spectra are obtained for triethyltin acetate $[(C_2H_5)_3Sn{-}O{-}CO]{-}CH_3$ [68] and the triethyltin polymer based on methylmethacrylate [65],

$$[(C_2H_5)_3\,Sn - O - CO] - C\begin{matrix} / CH_3 \\ \backslash\backslash CH_2 \end{matrix}$$

(see Fig. 13), in which no difference in the bonds occurs except far from the tin – outside the part in square brackets. The same spectra were also found for tetraphenyltin and triphenylstyryltin $(C_6H_5)_3Sn{-}C_6H_4{-}CH{=}CH_2$, in which one of the benzene rings in the molecule has a vinyl group in the para-position. In the general case, both the atoms far removed from the emitter or absorber and the functional groups (although more weakly) have

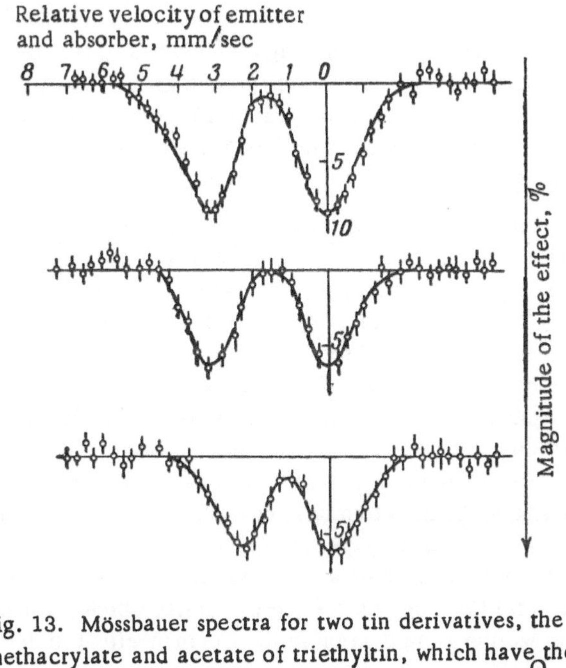

Fig. 13. Mössbauer spectra for two tin derivatives, the methylmethacrylate and acetate of triethyltin, which have the same chemical bonds nearest the tin: $(C_2H_5)_3\,Sn{-}O{-}C{\lt}^{O}$ (top – triethyltin methylmethacrylate; center – triethyltin acetate). The spectrum at the bottom is for an organotin compound in which the tin is bound not to ethyl but to phenyl radicals: $(C_6H_5)_3Sn{-}O{-}C{\lt}^{O}$ (triphenyltin methylmethacrylate). Temperature 77°K. Source – SnO_2.

an effect on the nature of the chemical bonds of the Mössbauer atoms themselves, and thus they produce a change in the chemical shifts and the quadrupole splitting of the spectrum lines or in the asymmetry of the two components of the splitting, which we will discuss in more detail below. An example of such an effect of more distant groups is provided by the difference, mentioned in our paper [68] and shown in Fig. 13, between the chemical shift and the quadrupole splitting of triethyltin, methylmethacrylate and triphenyltin. In both cases, the tin atom is bound to one oxygen and three carbon atoms, but the bond with the three carbon atoms in the aromatic groups has a somewhat smaller electric field gradient and a smaller electron cloud density in the region where the tin nucleus is located than is the case for the three bonds between the tin and the fatty ethyl groups.

A similar difference is apparently observed for derivatives of the type R_2SnCl_2, where R is aryl or alkyl, as shown below in the compilation of the data on tin (see Fig. 19). The reduction in electron cloud density in the region where the tin nuclei are located on going from fatty to aromatic substituents may be regarded as a consequence of the inductive electron attraction effect (I-effect), which is known to occur to a greater extent with aryl groups than with alkyl groups. The reason for this reduction may also be formation of dative π-bonds between the phenyl groups and the tin atom.

On the other hand, since in the examples given the tin is bound with (in addition to aryl and alkyl radicals) such strong electron acceptors as chlorine and oxygen, the four valence bonds are strongly nonequivalent in their ionic character, which causes quadrupole splitting to occur. On going from alkyls to aryls, i.e., to radicals which are more active as electron acceptors, the difference in ionic character of the bonds is somewhat smoothed out, so that the quadrupole splitting is reduced.

Thus, a study of Mössbauer spectra makes it possible to determine the character of the nearest chemical bonds of the absorber (or emitter) atoms, investigate the mutual effect of the bonds, and, apparently, the influence of such factors as inductive and mesomeric effects, and the inductomeric polarizability of the various near and more distant substituents on the electron shell structures of Mössbauer atoms. Thus, while the usual methods of finding the effect of these factors, for example, comparing the acidity of benzoic, acetic, and chloracetic acids, show that changing the electron acceptor properties of the end group affects the electron density on the other end of the molecule (for example in an acidic hydrogen atom), a comparison of the chemical shifts and the quadrupole splitting for tin in the examples given shows how the electric field in the center of the molecule depends on the nature of the end groups.

Possible Complications from a Preceding Conversion of the Emitting Nucleus

Since the Mössbauer emitting nucleus is formed as a product of a preceding beta decay (for example, $Os^{191} \rightarrow Ir^{191*}$, see Fig. 2b), electron capture (for example, $Co^{57} \rightarrow Fe^{57*}$, see Fig. 2c), or high-energy gamma-emission (for example, $Sn^{119m} \rightarrow Sn^{119*}$; see Fig. 2a), the chemical state of the emitter atoms is, in the general case, by no means unique. As a matter of fact, the "stirring up" of the electron shells that occurs during beta and gamma transitions produces multiple ionization of the emitting atoms, and as a result the atoms can be in the most diverse – including metastable – chemical states. Theoretical calculations of the relaxation time of these metastable states give a very small value as compared with the duration of the subsequent gamma transitions ($\sim 10^{-12}$ sec) [69]. However, various chemically stable states can also be formed in the emitter atoms, with different chemical shifts in the Mössbauer gamma transition energy, and as a result the form of the absorption spectra will be distorted. It was shown in a recent paper by Wertheim [69] that, when using a cobalt oxide CoO source instead of Co^{57}, the instant the following gamma emission occurs the Fe^{57} atoms are in both the Fe (Fe^{2+}) state, forming an oxide isomorphic with CoO, and the Fe··· (Fe^{3+}) state, which produces appreciable splitting of the Mössbauer line (see Fig. 14), as well as complication in the Zeeman-effect picture.

Thus, on the one hand, there is the problem of selecting an emitting material that will give the purest "single state" Mössbauer atoms, which is of special importance in chemical investigations. On the other hand, however, the Mössbauer effect presents obvious possibilities for making direct observations of the various chemical consequences of nuclear transformations, such as the formation, retardation, and annealing of all possible kinds of radiation defects.

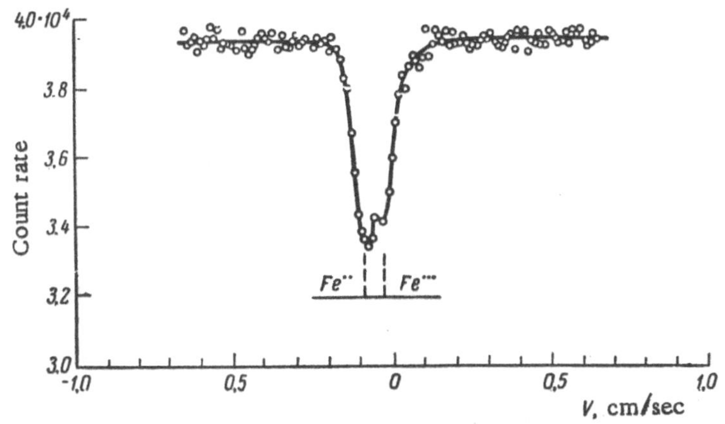

Fig. 14. Mössbauer spectrum splitting as a result of two chemical states of iron (Fe·· and Fe···) formed in K-electron capture by nuclei of Co^{57} in cobalt oxide. Absorber $K_3Fe(CN)_6$; temperature 298°K.

Asymmetric Doublet Splitting in Mössbauer Spectra

It may be seen from Fig. 14 that the presence of two chemical forms results in asymmetric splitting of the Mössbauer spectrum. It is perfectly natural to have asymmetry of this sort in all cases where the doublet comes from two different chemical states of the Mössbauer atoms, i.e., the doublet is formed by two singlet lines with different chemical shifts. Examples of such structures in which iron performs two functions, as determined by means of the Mössbauer effect [55, 70, 71], are provided by certain mixed oxides (for example, titanium–iron, or yttrium–iron garnets) in which the iron is nominally trivalent, but has a coordination number of 4 or 6 (tetrahedral and octahedral lattices). It is true that in these cases the form of the spectra is affected by the Zeeman and quadrupole splitting in addition to the double value of the chemical shifts. However, having asymmetric doublet splitting in the Mössbauer spectra is in itself no proof of the absence of quadrupole splitting or that the emitting or absorbing atoms are performing a double chemical function such as we had assumed, for example, at the start in the case of the organotin compounds of [68] or as was done in a recent paper [72] for hemin for the same reason.

In the paper already cited [69], interpreting the doublet asymmetry demonstrated above in Fig. 14 by asserting that there are two chemical forms of iron present is based entirely on denying from the very beginning, and completely without reflection, that it is possible to have asymmetric quadrupole splitting of Mössbauer lines.

Nevertheless, no longer speaking of the trivial asymmetry in the quadrupole splitting due to anisotropy of the test sample and some definite orientation with respect to the direction of the gamma quanta, we must recall here an important fact demonstrated in our papers [52, 73-75]: asymmetric quadrupole splitting of the Mössbauer spectra will be observed even in perfectly isotropic polycrystalline samples, if there are polycrystals present in which the Mössbauer effect is anisotropic [76-79]. Anisotropy of the Mössbauer effect in single crystals may lead to a condition where two singlet lines from two different chemical forms will be represented to a different degree, depending on the total anisotropy of the test sample or the angle that the anisotropic sample makes with the direction of the gamma quanta. In this case the doublet asymmetry will also depend on both of the above factors, so that the only rigorous proof of the trivial superposition of two peaks from different chemical forms suggested in a number of papers is that the asymmetric splitting is independent of anisotropy or orientation of the sample. We shall return to this question in Chapter VI.

CHAPTER V

PRINCIPAL RESULTS OF THE FIRST STUDIES OF THE MÖSSBAUER EFFECT IN IRON COMPOUNDS

The most complete Mössbauer effect studies made so far are those on iron and tin compounds using Co^{57} and Sn^{119m} sources. Before discussing the principal results obtained on iron compounds, we shall give a compilation of the data in Figs. 15 and 16. Since the literature references are only shown on the figures when different results were obtained for the same or similar compounds in different papers, we shall give the principal literature sources here.*

The spectra of iron in different metals and alloys were obtained in [55, 61, 80, 81]. Various iron salts were studied in [55], in the papers by de Benedetti et al. [83], and by Kerler and Neuwirth [83, 84]. In [55, 70, 71], studies were made of simple and mixed (for example, garnet) iron oxides. The nitride Fe_4N formed the subject of [85], and the two forms of the sulfide FeS_2 – pyrite and marcasite – were dealt with in the papers by Solomon [86, 87]. The spectra of rare earth ferrides were observed by Wertheim and Wernick in [88], UFe_2 in a paper by a group of Japanese authors [89] and other intermetallic iron compounds in [90]. The papers by Zahn, Kienle, and Eicher [91] and by Epstein [72] dealt with the Mössbauer spectra of ferrocene and of the ferricenium cation, while in [72] a study was also made of cyanide iron complexes (see also [55, 71]), as well as of carbonyl, phthalocyanine, hemin, acetylacetonate, and other complex iron compounds.

The complete data on chemical shifts relative to Fe^{57} in stainless steel are given in Fig. 15, while the information on quadrupole splitting (the distances between the two doublet peaks) is given in Fig. 16. The data are given for room temperature, but where a study was made of the temperature dependence of the chemical shift δ or the quadrupole splitting Δ in [82, 83, 88], vertical lines are drawn on the figures giving the upper and lower limits of the temperature range.

It may be seen from Fig. 16 that the quadrupole splitting sometimes increases by quite a bit when the temperature is lowered. As for the temperature dependence of the chemical shifts, it can be practically wholly explained, not by a change in $|\psi(0)|^2$, (i.e., by the actual chemical shift), but by the second-order Doppler effect mentioned above. Only in special cases (for example, ferricyanides) is it possible to have a change as a result of temperature changes in the 3d-electron screening effect [83].

It is obvious from Fig. 15 that, according to the magnitude of the chemical shift, inorganic iron compounds can be divided into three groups, corresponding to metallic, bivalent, and trivalent iron. Since the electron density at the nucleus is assumed to be greater in metallic iron than in the trivalent state, which corresponds to $\Delta R(Fe^{57}) < 0$ [50], the value of $|\psi(0)|^2$ is least for bivalent iron. The treatment of these results in [55], based on the calculations of Watson and Freeman [53, 54] leads to the conclusion that the electron configuration of metallic iron Fe^{57} in metals is nearly $3d^74s$, while trivalent and bivalent iron compounds have electron shells of the type $3d^54s^x$ and $3d^64s^x$, respectively, where x is the fraction of the 4s vacancies in the shells of

*Figures 15, 16, and 19 also give references to work, with which we had not become acquainted until after the manuscript had been submitted for publication. These papers ([122] et seq.) are given in the literature cited at the end of the book.

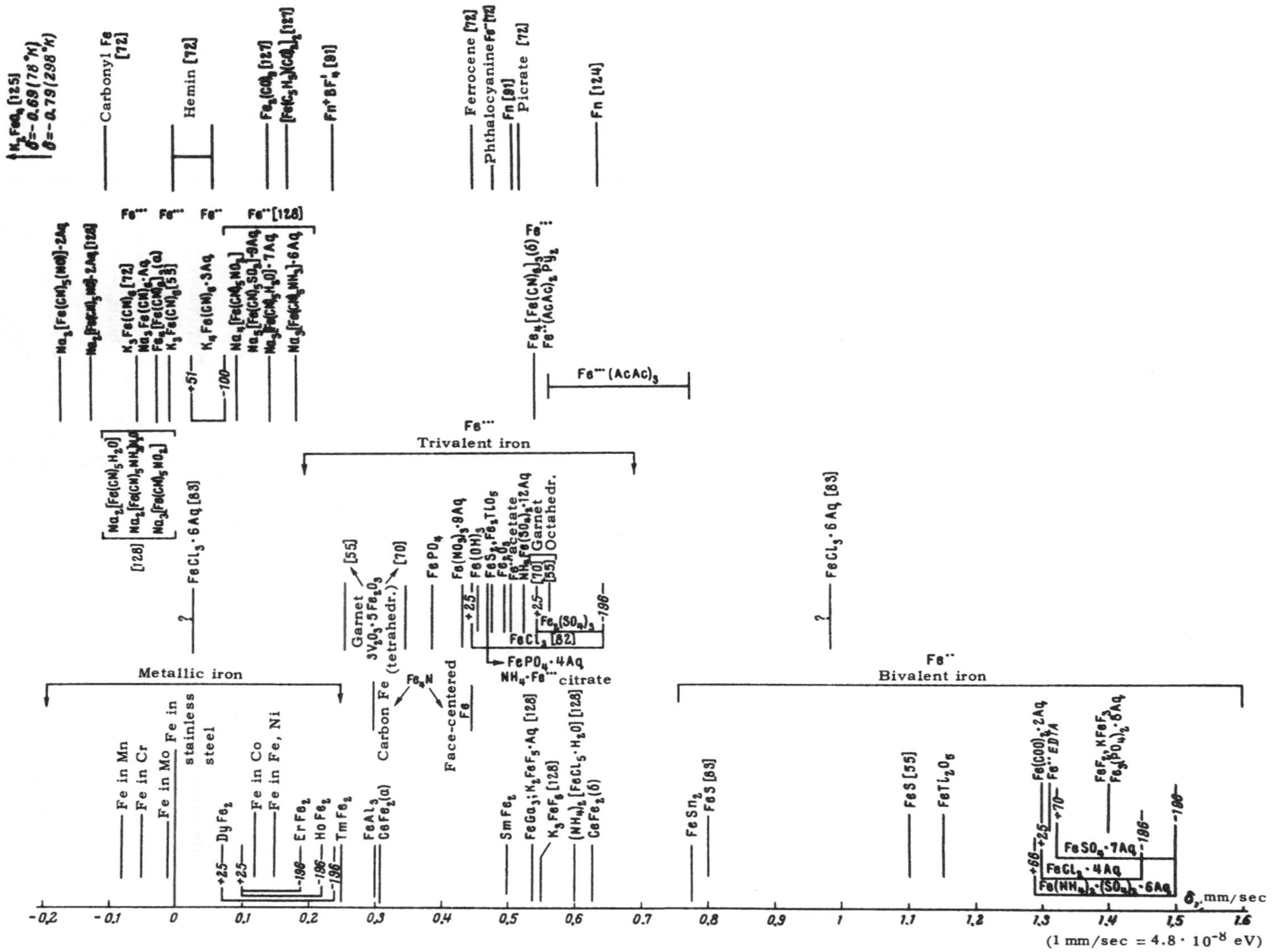

Fig. 15. Compilation of data on the chemical shifts in the Mössbauer spectra of iron compounds. The position of the line for iron in stainless steel is taken as zero.

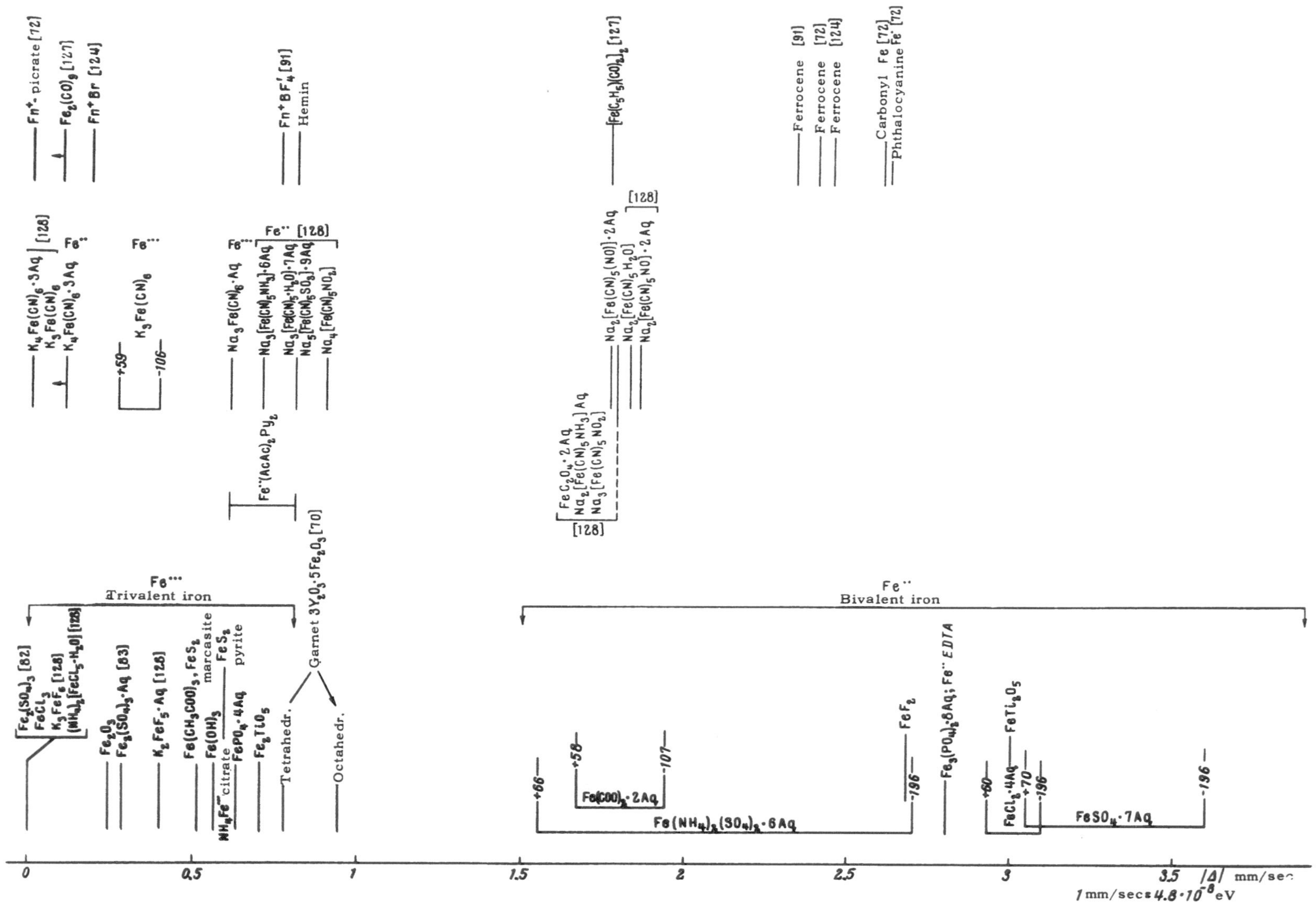

Fig. 16. Compilation of the data on the quadrupole splitting of lines in the Mössbauer spectra of iron compounds.

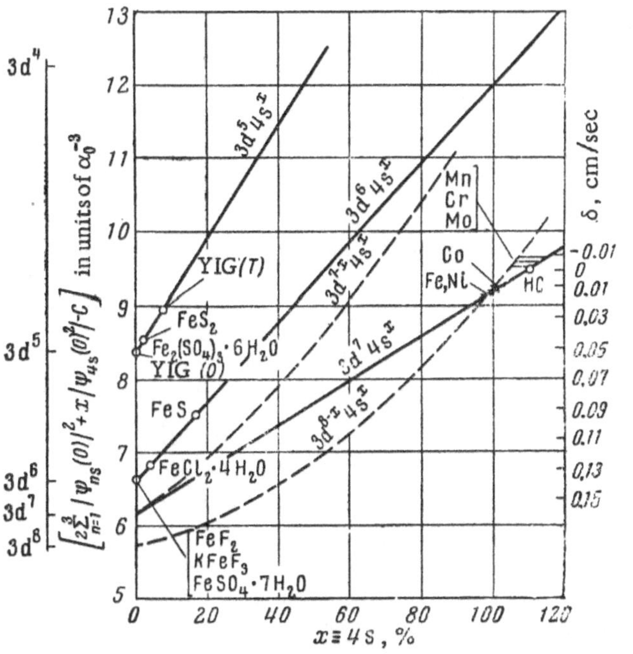

Fig. 17. Interpretation of the chemical shifts for iron compounds as given in [55]. C = 11.873 a_0^{-3}. The YIG are yttrium-iron garnets (T is the tetrahedral and O is the octahedral lattice).

the iron ions filled by electrons from the anions. These conclusions are illustrated in Fig. 17, wnere the left-hand ordinates are the calculated values [53, 54] of $2\sum_{i=1,2,3}' |\psi_{is}(0)|^2 - C$, i.e., the total densities of 1, 2, and 3s-electrons at the nucleus for different numbers (n = 4, 5, 6, 7, 8) of 3d-electrons. The straight lines $3d^n4s^x$ for n = 5, 6, 7 give the additions to $2\sum_{i=1,2,3} |\psi_{is}(0)|^2$ with filled $1s^22s^2p^63s^2p^6d^n$ configurations from 4s-electrons. The calculation is made on the assumption that these additions are simply $x \cdot |\psi_{4s}(0)|^2$, where the magnitude of $|\psi_{4s}(0)|^2$ is found by the Fermi–Segrè–Goudsmit method [92, 93] (here no account is taken of 4s-electron screening of the internal s-shells). In choosing the scale of the chemical shifts and in defining the value of ΔR/R, it is assumed that the compounds having the largest values of δ in a given valence group, namely $Fe_2(SO_4)_3 \cdot 6H_2O$ and the octahedral lattice of the garnet $3Y_2O_3 \cdot 5Fe_2O_3$ for $Fe^{\cdot\cdot\cdot}$, and FeF_2, $KFeF_3$, and $FeSO_4 \cdot 7H_2O$ for $Fe^{\cdot\cdot}$, are the "standards," having pure $3d^5$ and $3d^6$ electron configurations. These calculations give the value $\Delta R/R = -1.8 \cdot 10^{-3}$ and the values of x in the $3d^n4s^x$ configurations for all the other compounds in the $Fe^{\cdot\cdot\cdot}$ and $Fe^{\cdot\cdot}$ series. The data for ferro- and ferricyanides [55] are not being analyzed. It must, however, be noted that in interpreting the Mössbauer spectra it is necessary to take account of not only formal valences, but also the true picture of the distribution of the electron cloud, for a description of which it is important to have the values of the electric charges (η) of the various atoms in chemical compounds, obtained by a study of the fine structure of the edges of the x-ray K-absorption spectra [94, 95, 96]. The idea on which these studies is based may be seen from the simple example of the x-ray spectra of atoms of inert gases, say, argon. If a 1s-electron is removed from the K-shell of an argon atom in the free state, a continuous x-ray absorption spectrum will be produced. However, in the vicinity of the corresponding ionization potential, the absorption spectrum shows a fine structure corresponding to the various possible transitions with $\Delta l = 1$, i.e., transitions from the 1s state to the free p-levels 4p, 5p, 6p, etc. Removing one electron from the K-shell produces a "hole" in the immediate vicinity of the nucleus, i.e., the nuclear charge increases, as it were, by unity, so that instead of argon we get something like potassium. As a matter of fact, the distances between the fine-structure lines corresponding to the transitions 1s → 4p, 5p, 6p, etc. correspond to the energy difference of the optical terms in potassium.

Knowing the distance between several of the fine-structure lines at the edges of the x-ray absorption spectra, the intensity ratio of the lines, and their total contribution in comparison to the continuous spectrum, we may determine such parameters as the effective nuclear charge, the effective value of the principal quantum number, and, of particular importance, the effective charge of the ion, in the field of which the electron that has jumped from the K-shell to one of the distant orbits is located.

In our example, this charge, naturally, proves to be unity. But in cases where the original atom was not in itself electrically neutral, but had some effective charge η, the charge on the ionic residue after the K-electron jump proves to be equal to $+(1 + \eta)$.

Thus, a study of the fine structure of the edges of the x-ray absorption spectra makes it possible in principle to determine the effective charges η of both free atoms and those forming parts of molecules, without involving the valence electrons, as would be the case if optical spectra were used for the purpose.

For a free neutral atom, $\eta = 0$, but the value of the positive or negative effective charge of a bound atom is characteristic of the reduction or increase in the effective number of electrons in the atom when the chemical bonds are formed. If N is the number of valence electrons in the free neutral atom, then $A = N - \eta$ is the effective number of valence electrons which the atom in question has at its disposal in the chemical compound. With purely ionic bonds, the values of the positive or negative electric charge of the ion I^Z and its effective charge η would simply be the same: $Z = \eta$, i.e., in the present case $A = N - Z$. For the more complex bonds described within the framework of the theory of molecular orbitals, the effective number of valence electrons is $A = \Sigma a_i^2 = \Sigma a_s^2 + \Sigma a_p^2 + \Sigma a_d^2 + \ldots$, where the a_i^2 terms give the contribution made by the atomic orbitals of a given atom to the molecular orbitals (see p. 31). In view of what has been said, the chemical state of the ion I^Z should not be designated by the atomic term corresponding to the given value of Z, if the effective charge $\eta \neq Z$, since this term will then describe neither the purely ionic state with the charge η, nor the bonds characterized by the molecular orbitals. This is one of the results of experiments made to determine the effective charges of bound atoms.

Calculating the effective radius of the hydrogenlike system consisting of an ionic residue and a K-electron that has jumped to a more distant orbit gives a value of about 1 A, i.e., about the same as the interatomic distances in the molecule. Thus, when we are talking not about free atoms but about those bound in molecules, the electron removed from the K-shell is not just in the field of the ionic residue itself but in that of the other atoms; this can, generally speaking, distort the effective charge values found as described above.

Leaving aside, however, any possible critical comments on the x-ray determination of η, we will now consider an alternative method of treating the data on the chemical shifts in the Mössbauer spectra of iron, based on the values given in [95] for the effective charges of the iron atoms in various compounds. For $Fe^{\cdot\cdot}$ salts, $\eta = +1.9$; for $K_3Fe(CN)_6$, $\eta = +1.0$; for the carbonyl $Fe(CO)_5$ and ferrocene $Fe(C_5H_5)_2$, $\eta = +0.4$, and the values of η for ferrocene (Fn) and the ferricenium cation (Fn^+) are almost the same [96]. In view of the analogy between iron and cobalt, the value $\eta = +1.2$ should be taken for $Fe^{\cdot\cdot\cdot}$, i.e., it should be assumed that the effective charge for trivalent iron is not greater, but less than that of bivalent iron.

It is interesting that the same conclusion may be arrived at on the basis of Fig. 15, if we set aside the question of screening of 4s-electrons by the 3d-shell. As a matter of fact, the values of the chemical shifts for $Fe^{\cdot\cdot\cdot}$ salts lie between those for metallic iron and $Fe^{\cdot\cdot}$ salts.

Assuming now that the effective number of valence electrons is $A = 8 - \eta$ and that $A = 8$ for metallic iron, we can interpret the chemical shift data for iron in the light of Watson and Freemen's calculations [53, 54] in the same way as is done in Fig. 18. Then we must assume a structure for metallic iron, which is close to $3d^6 4s^2$ (and not $3d^7 4s$) for $Fe^{\cdot\cdot\cdot}$-$3d^{6.8-x}4s^x$, where x = 1-1.25 – and for $Fe^{\cdot\cdot}$ – $3d^{6.1-x}4s^x$, where x = 0-0.35. In the same way, for Fe^{57} we obtain, instead of the value $-1.8 \cdot 10^{-3}$ [55]

$$\frac{\Delta R}{R} = -5 \cdot 10^{-4}.$$

The disagreement between the true number of electrons in the iron atoms and their formal valence state shows up with particular clarity in the case of potassium ferrate K_2FeO_4 [125], in which the iron is formally hexavalent

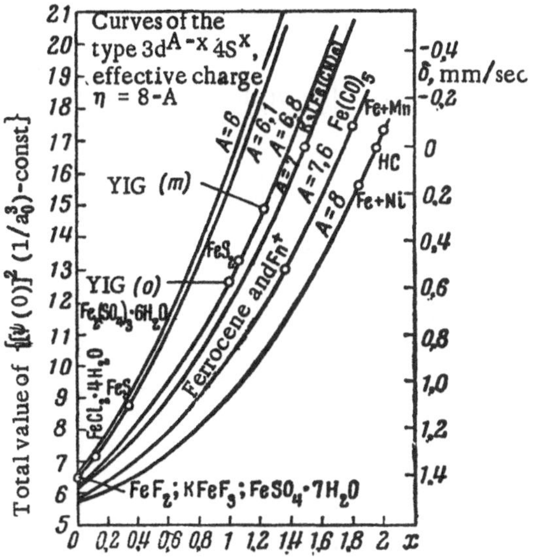

Fig. 18. Alternative interpretation of the chemical shifts for iron compounds. Includes the effective atomic charges (but without the sp^3d^2 hybridization for complex compounds). The constant (ordinate axis) is, as in Fig. 17, C = 11.873 a_0^{-3}.

($3d^2$), while the chemical shift with respect to stainless steel is δ = −0.69 (78°K) −0.79 (298°K) mm/sec (this shift is not given in Fig. 15 since it is off the scale to the far left).

The authors of [125] themselves point out that the observed shift here corresponds to the $3d^{4.3}$ configuration (see Fig. 17), from which we conclude that in K_2FeO_4 the effective charge of the iron atom cannot exceed η = +3.7. There is considerable interest in making a direct determination of η_{Fe} in compounds of ferrate type.

Figure 18 also gives points for the iron complex compounds, corresponding to the configurations $3d^{5.5}4s^{1.5}$ for $K_3Fe(CN)_6$, about $3d^{6.2}4s^{1.4}$ for ferrocene and ferricenium, and $3d^{5.8}4s^{1.2}$ for carbonyl. However, this would be too rough an interpretation, since it does not include sp^3d^2 hybridization, which shows up when iron coordination bonds are formed in complex compounds. If some of the 3d-electrons go to 4p-orbits $|\psi_{1,2,3s}(0)|^2$ would, as may be seen from Fig. 17, increase, i.e., the positive chemical shifts would be reduced, as is observed in all complex compounds. It is now easily seen that, having assumed a certain number of electrons pass to the 4p-orbit, we cannot reconcile this assumption with the observed values of the chemical shifts by considering 3d-electrons alone, and it is necessary to reduce the number of both 3d- and 4s-electrons at the same time (the curves in Fig. Fig. 18 shifts to the left and up). The possibility of having the electrons arranged in different ways in the 3d- and 4sp-orbits permits us, in principle, to explain simply why the values of δ are very nearly the same for different complexes in which the valence of the iron is nominally different. For the explanations to be at least approximately correct, we must, however, keep in mind both the chemical shifts and the effective charges, as well as the quadrupole splitting of the Mössbauer spectra. It may be seen from Fig. 16 that the splitting is considerably greater for Fe·· than for Fe··· salts, which is usually accounted for by the spherical symmetry of the fundamental $^6S_{5/2}$ term in contrast to Fe·· ($^5D_4$). However, this explanation contradicts the data given in [95, 96] on the effective charges of the ion in Fe··· salts. Further, the temperature dependence of the quadrupole splitting is greater for Fe·· than for Fe··· salts, which is accounted for in [82] by thermal excitation of the higher-lying Fe·· terms.

The specific effect of the structure of iron cyanide complexes, which makes them different from ordinary Fe··· and Fe·· salts, also shows up in the values of the quadrupole splitting shown in Fig. 16. Red potassium ferricyanide, like the usual Fe··· salts, shows rather small quadrupole splitting, and yellow potassium ferrocyanide shows no splitting at all (Δ < 0.1 mm/sec), while for ordinary Fe·· salts and sodium nitroprusside the distance between the two peaks is greater than 1.5 mm/sec. Nearly the same as the cyanide complexes of iron both in the magnitude of the chemical shift and in the quadrupole splitting of the Mössbauer spectra is [72] hemin (ferrihem), a complex of oxidized, trivalent iron with protoporphyrin. At the same time, a structure very similar to porphyrins – phthalocyanine Fe·· – shows values of chemical shift and quadrupole splitting considerably greater than for hemin. It is interesting, finally, to note that the dipyridine-acetylacetonate of bivalent iron [Fe··$(AcAc)_2(C_5H_5N)_2$], in both the value of δ, and in the quadrupole splitting, falls in the range of trivalent iron salts [72].

An explanation of the data on quadrupole splitting of lines in cyanide complexes of iron is given in [83]. These complexes are low-spin – here six d-electrons of Fe·· or 5 electrons of Fe··· occupy only three out of the five 3d-orbits of iron. The other two orbits take part in forming six hybridized $3d^2\,4sp^3$ bonds of

TABLE II

Symmetry of MO of complex	Ferrocene Fn [97]		Ferricenium Fn^+ [98]	
	form of aψ	no. of electrons in MO	form of aψ	no. of electrons in MO
A_{1g}	0.49s 1.00 d_{z^2}	2 2	0.49s 1.00 d_{z^2}	2 2
A_{1u}	0.10 p_z	2	0.09 p_z	2
E_{1u}	0.59 p_x 0.59 p_y	4	0.60 p_x 0.60 p_y	4
E_{1g}	0.37 d_{xz} 0.37 d_{yz}	4	0.45 d_{xz} 0.45 d_{yz}	4
E_{2g}	0.85 $d_{x^2-y^2}$ 0.85 d_{xy}	4	0.94 $d_{x^2-y^2}$ 0.94 d_{xy}	3

Fe with CN^-. In diamagnetic ferrocyanide $[Fe^{II}(CN)_6]^{IIII}$ there are six equivalent bonds, resulting in a spherially symmetric charge distribution and absence of quadrupole splitting. The presence of weak splitting for paramagnetic ferrocyanide $[Fe^{III}(CN)_6]^{III}$ [monoclinic crystals of $K_3Fe(CN)_6$] is the result of the appearance of a "hole" in the 3d-shell on passing from $Fe^{\cdot\cdot}$ to $Fe^{\cdot\cdot\cdot}$. Finally, the relatively strong splitting for nitroprusside is explained by the disturbance of the spherically symmetric charge distribution as a result of the nonequivalence of the bond of Fe with NO^+ and the five other bonds of Fe with CN^-.

This explanation seems more natural than that advanced in [72], which assumes that in nitroprusside the ligand is NO and not NO^+ and, accordingly, that the iron in the complex is tetravalent.

We turn now to ferrocene. Most characteristic in this case is the great difference between the quadrupole level splitting in ferrocene itself, where the splitting is very strong (Δ = 2.33-2.4 mm/sec [91, 72]), and in ferrocene compounds, where the splitting is much weaker and, moreover, apparently depends on the nature of the anion (Δ = 0 for the picrate of Fn^+ [72], and Δ = 0.75 mm/sec for $Fn^+BF_4^-$ [91]). However, the chemical shifts lie in the range of values characteristic of $Fe^{\cdot\cdot\cdot}$ and are nearly the same for ferrocene (0.46 mm/sec [72] and 0.52 mm/sec [91]) and Fn^+-picrate (0.53 mm/sec), and are somewhat less (0.25 mm/sec) for $Fn^+BF_4^-$.

An interpretation of the quadrupole splitting data has been advanced in [91]. As we know, the $Fe(C_5H_5)_2$ contains 18 valence electrons (8 from iron and 5 from each ring), which corresponds to completely filled iron shells. Here, there would be no quadrupole splitting at all if the σ-, π-, and δ-bands in the group of d-electrons (i.e., the group of electrons with magnetic quantum numbers m_l = 0, ±1, and ± 2) had the same values $\overline{r^3}$ for the mean cube of the distances from the iron nucleus, since the contribution of one d-electron of each band to the value of Δ is proportional to $(m_l^2-2)/\overline{r^3}_{ml}$, and $2(0^2-2)+4(1^2-2)+4(2^2-2) = 0$. A stronger bond with the ligands corresponds to an increase in Δ, and thus the authors of [91] account for the observed quadrupole splitting (assuming that $\Delta > 0$, i.e., the $\pm 1/2 \to \pm 3/2$ transition energy is greater than that for a $\pm 1/2 \to \pm 1/2$ transition) by saying that the δ-electrons of the iron are more weakly represented in the vicinity of the ligands than the σ- and π-electrons, and hence they have a larger value for the factor $1/\overline{r^3}_{ml}$. The reduction in the value of Δ on going from ferrocene to the ferricenium salt $[Fe(C_5H_5)_2]^+BF_4^-$ is explained by the authors of [91] by asserting that, when ferrocene is ionized, a d-electron is taken out of a δ-band, in which the electron is "closest" to the iron and thus relatively weakly bound with the ligand rings, so as to make a positive contribution to the quadrupole splitting.

In interpreting the data of [91] on ferrocene and ferricenium, it is convenient to use the results of the only quantitative calculations of their kind on these complexes — made by E. M. Shustorovich and M. E. Dyatkina

[97, 98] using the self-consistent molecular orbit (MO) method. The valence shells of the complexes consist of nine molecular binding orbits of various symmetries, of the form $\varphi = a\psi + b\chi$, where the ψ's are the atomic orbits of iron, the χ's are the molecular orbits of the ligand rings, and a and b are numerical coefficients, with $a^2 + b^2 = 1$. The results of the calculations are given in Table II, where values of a are given for each of the molecular orbits of the complexes. It is obvious that the density of the 4s-electron cloud (on account of the A_{1g} orbit) is practically the same in Fn and Fn+ $(2 \cdot 0.49^2 \approx 0.5)$; this means that the chemical shifts are about the same in both complexes. The reduction in the quadrupole splitting in ferricenium not only comes from the reduction in the E_{2g} contribution $(3 \cdot 0.94^2 - 4 \cdot 0.85^2 = -0.26)$, but also from the increase in the E_{1g} contribution $(4 \cdot 0.45^2 - 4 \cdot 0.37^2 = 0.24)$ in Fn^+, since E_{2g} gives positive $(m_l^2 - 2 = +2)$ splitting and E_{1g} gives negative $(m_l^2 - 2 = -1)$ splitting (the A_{1g} contribution from d-electrons with $m_l = 0$ remains unchanged in this case).

It may be seen from Table II that the E_{2g}-electron removed belongs by more than 70% to the atomic orbit of iron, i.e., the factor $1/r^3$ for this electron may quite naturally be assumed greater than for the electrons making a negative contribution to the quadrupole splitting. However, the rearrangement occurring when the E_{2g}-electron is removed not only leads (in two ways), as has already been said, to a reduction in quadrupole splitting, but also serves to maintain the effective charge of the iron atom. According to Shustorovich and Dyatkina [97, 98], the distribution of the total number of electrons in the iron atoms in ferrocene (7.3) and ferricinium (7.4) has the form $3d^{5.4}4s^{0.5}p^{1.4}$ ($p^{1.5}$ for ferricinium). It is interesting that if we neglect the effect of the 4p-electrons on $|\psi_{1,2,3s}(0)|^2$ and consider only the $3d^{5.4}4s^{0.5}$ structure, a calculation of $|\psi(0)|^2$ for this structure by the Watson–Freeman [53, 54] method (in the sense of Fig. 18) gives beautiful agreement with the chemical shifts observed for ferrocene and ferricenium. This is an additional evidence for the correctness of the theoretical calculation of the ferrocene structure given in [97, 98].

In concluding our brief analysis of the data on iron, let us stop to consider the values of the chemical shifts for $FeCl_3 \cdot 6H_2O$, which, as may be seen from Fig. 15, fall far out of the series of trivalent iron salts. The value given for this salt in [83] is $\delta = +0.98 \pm 0.03$ mm/sec, and a "secondary" line, with $\delta = +0.03 \pm 0.05$ mm/sec, is also mentioned. The reason for distinguishing between a main and a secondary line is that the difference in resonance effect between them is more than a factor of 1½ (15.9% in the main, and 10.2% in the secondary line), i.e., asymmetric doublet splitting is observed which the authors of [83] were not able to interpret as quadrupole splitting. However, in the light of our papers already mentioned [52, 73-75], in which the asymmetric quadrupole splitting for isotropic polycrystalline samples is accounted for by the anisotropy in the Mössbauer effect in the single crystals themselves, the case of $FeCl_3 \cdot 6H_2O$ finds a natural explanation – the chemical shift is $\delta = +0.50$ mm/sec, which is the same as the data of [82] for anhydrous $FeCl_3$, but the value of $\Delta = 0.95$ mm/sec also falls in the range for trivalent iron salts, although at the very edge of the range (the difference between Δ for iron chloride with water of crystallization and for the anhydrous chloride, where $\Delta = 0$, is still in need of further clarification).

The local magnetic fields in the region where the iron nuclei are located, measured in a number of papers with the aid of the Mössbauer effect, also depend on the chemical state of the iron. In metallic iron [80], cobalt, nickel [61], and copper–nickel alloy [81] the local magnetic fields in the region where the iron nuclei are located are approximately the same and are about 300,000 Oe. Accordingly, these fields are determined by the electron shell structure of the iron itself and occur, for example, as a result of contact Fermi interaction through exchange polarization of the internal s-electrons with the unfilled 3d-shell of the iron atoms. In oxides containing trivalent iron, as was first shown in [50], even stronger fields, of nearly 500,000-550,000 Oe, are observed, and they are almost the same in all such systems investigated (for example [71]), with the exception of the tetrahedral lattices of yttrium iron garnets, where H = 390,000 Oe [50]. Thus, it is possible even in these cases to speak of definite local magnetic fields, inherent in the actual electron shell structure of $Fe^{\cdot\cdot\cdot}$. Finally, in the ferrides of a number of rare earth elements (Sm, Gd, Dy, Ho, Er, and Tm) of the type $SmFe_2$, the magnetic field in the vicinity of the iron nucleus is also almost always the same and equal to 230,000 Oe, and

* The effective charges of the iron atoms used in [97, 98] are: $\eta = +0.7$ for ferrocene and +0.6 for ferricenium, while on pages 28-29 we mentioned another value: $\eta = +0.4$ [95]. The differences in the values of η given are small and unimportant, and the only thing that is important to us just now is that the effective charge of the iron atoms is the same in both ferrocene and ferricenium.

a different value of H = 310,000 Oe is observed only in $CeFe_2$ [88]. Accordingly, the iron atoms have approximately the same electron configuration in all these compounds (and, moreover, judging from the chemical shifts – see Fig. 15 – it is nearly the same as that in metallic iron), while the contribution from polarization of conductance electrons is very small or is determined by interaction with the d-electrons of the iron. In the case of $CeFe_2$, it is assumed in [88] that 4f-electrons of cerium are transferred to the d-band of iron and, judging from the chemical shifts, they are not localized at any definite iron atoms. Of the other ferrides, we can mention the uranium ferride UFe_2 investigated in [89], for which H = 20,000 Oe at room temperature and 65,000 Oe at T = −195°C. The most definite character so far has been shown by the data on the local magnetic fields in the region where bivalent iron nuclei are located, which apparently depend on the environment of the iron atoms, and are recounted in the paper by Wertheim [69] already mentioned, from which Fig. 13 was taken. It was also shown that for FeF_2, H = 340,000 Oe, and for Fe (II) in Fe_3O_4, H = 450,000 Oe. In the same paper [69], where two forms of iron $Fe^{\cdot\cdot}$ and $Fe^{\cdot\cdot\cdot}$ were formed in the beta decay of Co^{57} in cobalt oxide CoO, which is an antiferromagnetic cubic structure with a Néel temperature of +18°C, it was found that the field in the region where the bivalent iron nuclei are located is 200,000 Oe at quite low temperatures, while for $Fe^{\cdot\cdot\cdot}$ the characteristic value of H = 560,000 Oe is observed even here. Figure 14 given above was obtained in [69] at t = 25°C, and there is therefore no Zeeman splitting.

In summary, it may be said that, unlike the chemical shifts and the quadrupole splitting, these not too numerous data on the magnetic splitting of the Mössbauer spectra of iron have not yet received any quantitative interpretation from the point of view of the electron shell structure of iron.

CHAPTER VI

PRINCIPAL RESULTS OF THE FIRST STUDIES OF THE MOSSBAUER EFFECT IN TIN COMPOUNDS

We turn now to the data for tin. A compilation similar to Figs. 15 and 16 of the results on chemical shifts and quadrupole splitting is shown schematically in Fig. 19. The data on inorganic tin compounds are taken from our paper [52] and the papers by V. S. Shpinel' [51, 99] and Boyle and co-workers [67], as well as from the compilation given in [126]. The values of δ and Δ for organotin compounds were obtained in our papers [52, 66, 68] (for the radicals ethyl C_2H_5 = Et, propyl C_3H_7 = Pr, and phenyl C_6H_5 = Ph), as well as in the papers by L. S. Polak and V. S. Shpinel' and co-workers [105, 123] (for the butyl radical C_4H_9 = Bu). Figure 19 does not contain all the information on the chemical shifts and quadrupole splittings in the spectra of organotin compounds, but the data given are sufficient to sketch out the values of these quentities characteristic of organotin compounds. The literature references are given in Fig. 19 only in those cases where the data are in substantial contradiction, as is the case, for example, for the identical bivalent tin compounds investigated in [51, 99, 67, 126].

If a tin compound is shown only in the graph of chemical shifts but not in the quadrupole splitting graph (lower), then $\Delta = 0$ for the compound. As far as the basic features of the data shown in Fig. 19 are concerned, as a rule, all the quadrivalent tin compounds fall in the range of negative chemical shifts with respect to β-Sn, while bivalent tin compounds are in the range $\delta > 0$. With all the valence electrons present in the tin atom, it is natural to assume a completely hybridized structure of the type $5s5p^3$. The wave function of each of the four hybridized orbits is of the form $\psi_{hybrid} = (1/2)\psi_s + (\sqrt{3}/2)\,\psi_{pz}$, i.e., the contribution from each such orbit to the electron density at the region where the nucleus is located is equal to $1/4\ |\psi_{5s}(0)|^2$. Accordingly, neglecting the perturbing effect of the 5p-electrons on $\sum_{i=1...4} |\psi_{is}(0)|^2$, we may conclude that with four covalent bonds, the electron density at r = 0 corresponds approximately to the presence of one s-electron. With completely ionized $Sn^{::}$, this electron is taken away; what remains is the structure of completely filled n = 1, 2, 3, and 4 spd-shells. Accordingly, the value of $|\psi(0)|^2$ falls below its value for entirely covalent bonds. On the other hand, in completely ionized bivalent tin, the hybridization is destroyed, and two 5p-electrons are taken away, leaving the two 5s-electrons, i.e., the value of $|\psi(0)|^2$ becomes greater than its value for completely covalent compounds, and also greater than its value for completely ionic compounds as quadrivalent tin. Using a similar argument, we not only find a positive sign for $\frac{\Delta R}{R}(Sn^{119*})$ without difficulty, but also arrive at the conclusion that strengthening the ionic character of the quadrivalent tin bonds corresponds to increasing the chemical shift in the negative direction (from SnI_4 to SnF_4), while, for bivalent tin, strengthening the ionic character of the bonds increases the chemical shift in the positive direction – from SnO to $SnCl_2$. An analysis of the data on bivalent tin compounds has so far been rendered difficult by the fact that the data are obviously in contradiction. It may be seen from Fig. 19 how much difference there is between the chemical shifts from SnF_2, $SnCl_2$, and $SnCl_2 \cdot 2H_2O$ (and in the last case, in the quadrupole splitting) given in the papers by V. S. Shpinel' [51, 99], Boyle and co-workers [67], and Kistner [126].

The contradiction is particularly striking between the data of [99] and [126] on the chemical shift for $SnBr_2$. The value of δ given in [99] (+0.2 mm/sec with respect to SnO_2) differs by 3.7 mm/sec from the value in [126] and in general departs sharply from all the known data on bivalent tin compounds. Apparently, the authors of [99] took some quadrivalent tin compound to be $SnBr_2$. As far as quadrivalent tin is concerned, we can already draw a number ot qualitative and even semiquantitative conclusions.

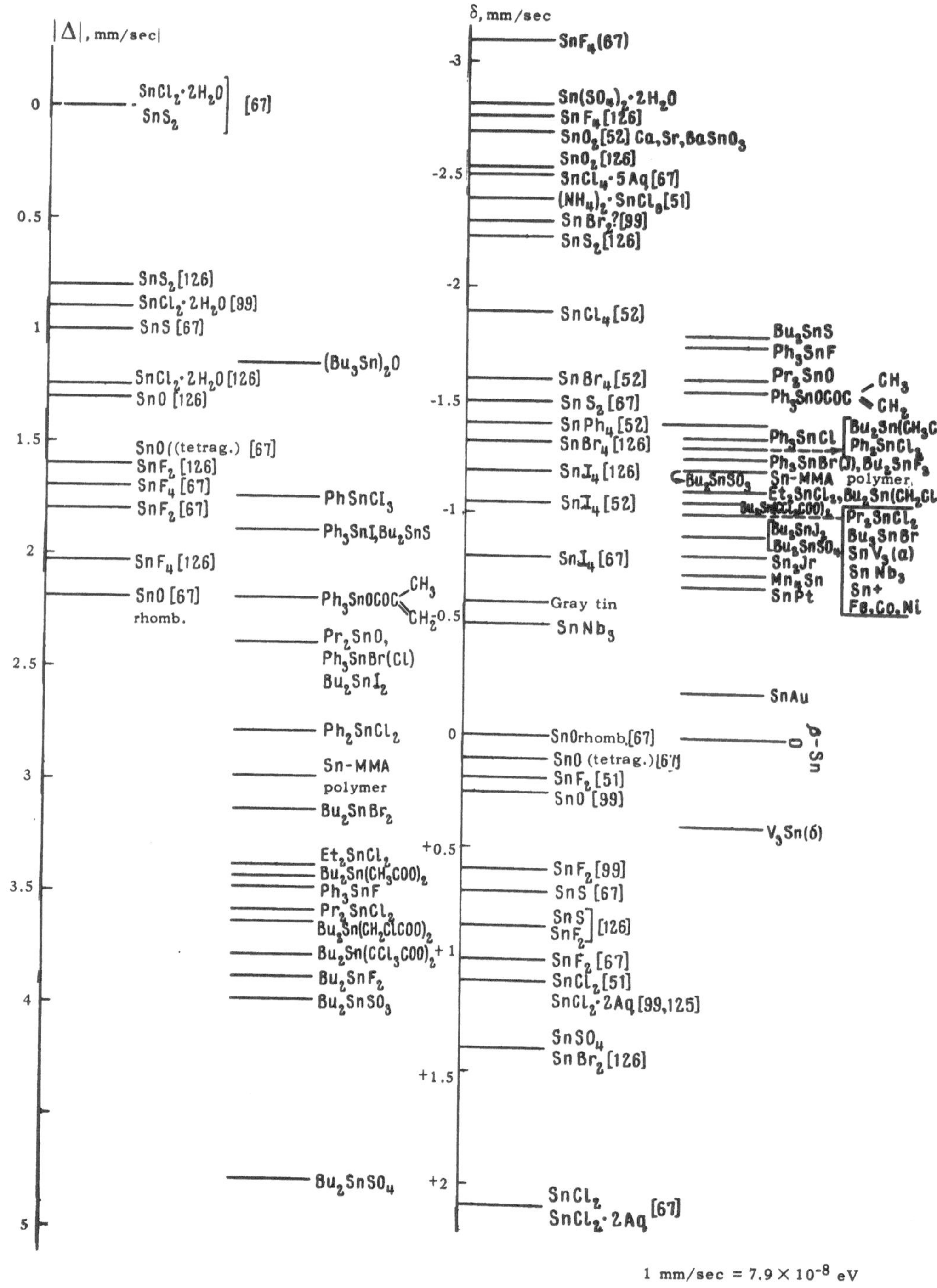

Fig. 19. Compilation of the data on chemical shifts (with respect to β-Sn) and the quadrupole splitting of the lines in the Mössbauer spectra of tin compounds. Ph – phenyl C_6H_5; Et – ethyl C_2H_5; Pr – propyl C_3H_7; Bu – butyl C_4H_9.

A calculation of $|\psi(0)|^2$ for a single 5s-electron in the field of the skeleton consisting of the tin nucleus and the electrons in the completely filled 1, 2, 3, and 4spd-shells, made by the Fermi–Segrè method [92], gives the value $1.5 \cdot 10^{26}$ cm^{-3} [67]. There are two more possible ways of making the calculation. Boyle et al. [67] assume that SnF_4 is an example of a completely ionized compound of quadrivalent tin, and $SnCl_2$ of bivalent tin. Then these two compounds differ by two 5s-electrons, and the chemical shift between them is equal to 5.2 mm/sec = $4.1 \cdot 10^{-7}$ eV.

According to (26) and (27), the chemical shift for Sn^{119} is equal to (in eV)

$$\delta_{Sn} = 1.55 \cdot 10^{-29} \frac{\Delta R}{R} \{|\psi(0)|^2_a - |\psi(0)|^2_{em}\}. \quad (30)$$

In [67] a value is given of $\Delta R/R = 1.1 \cdot 10^{-4}$, which corresponds to a difference in $|\psi(0)|^2$ between $SnCl_2$ and SnF_4 of $2.4 \cdot 10^{26}$ cm^{-3}. Although this value is not given directly in [67], nevertheless, judging from the text, it was found from including mutual screening of the two 5s-electrons and is thus less than $2\,|\psi_{5s}(0)|^2 = 3 \cdot 10^{26}$ cm^{-3}. The weak point in this calculation is the assumption of complete ionization in SnF_4 and $SnCl_2$. Accordingly in our work [52] we first of all compared the value of the chemical shifts in quadrivalent tin compounds of the type SnX_4 with the electronegativity of the X-atoms and with the degree of ionicity of the bond found for Sn–Hal bonds by independent methods (for example, the NQR method [100], or from refraction and dielectric constant studies [101]). The result of the comparison, shown in Fig. 20, enables us to find the value of δ for an extrapolated 100% ionic bond in SnX_4, giving $\delta = -(5.6 \pm 0.5)$ mm/sec $= -(4.4 \pm 0.4) \cdot 10^{-7}$ eV, which is 2.5 mm/sec greater than the value of δ for SnF_4. Note that even the point $\delta = 0$ for β-Sn itself fits well on the general linear curve given in Fig. 20 for the chemical shift as a function of the electronegativity of X (the Sn–Sn bond in this case). Making use of the fact that for completely covalent and completely ionic quadrivalent tin bonds the difference in $|\psi(0)|^2$ is determined by one 5s-electron, since $5s5p^3$ hybridization, as we have seen, does not change matters, we obtain from (30)

$$\frac{\Delta R}{R} = \frac{(4{,}4 \pm 0{,}4) \cdot 10^{-7}}{1.55 \cdot 10^{-29} \cdot 1{,}5 \cdot 10^{26}} = (1.9 \pm 0.17) \cdot 10^{-4},$$

after which we can make a direct determination of the contribution to $|\psi_{5s}(0)|^2$ in the other tin compounds from the value of the chemical shift. There is particular interest in all possible organotin compounds having strong quadrupole splitting almost up to $\Delta = 5$ mm/sec (which is in general absent in all quadrivalent tin compounds except SnF_4) and having chemical shifts in the range (–0.9) –(–1.8) mm/sec, i.e., in the vicinity of the values of δ for SnR_4, where R is an organic radical.

A comparison between the values of quadrupole splitting for tin compounds and the Townes–Dailey calculations [56] of the field gradients q can so far only be made on the basis of a very rough approximation for the quadrupole moment of Sn^{119*}, $Q = 8 \cdot 10^{-26}$ cm^2 [67]. These calculations give a distance between peaks of Δ = 4.6 mm/sec for a pure p_z electron and $\Delta = 3.5 \cdot x$ mm sec if, with complete $5s5p^3$-hybridization, one of the quadrivalent tin bonds is of a partly ionic character (by the fraction x). With complete $5s5p^35d^2$ hybridization, appreciable quadrupole splitting is obtained for the configurations

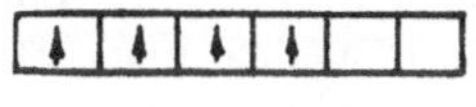 ($\Delta \gtrsim 4.6$ mm/sec) or 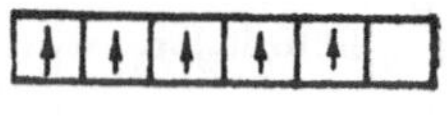 ,

where one of the six bonds is completely ionic (Δ = 2.4 mm/sec). For organotin compounds it is necessary to assume that there are four hybridized $5s5p^3$ bonds, which differ in degree of ionic character so as to produce inhomogeneity in the electric field. In the series Ph_3SnHal, where Hal = F, Cl, Br, or I, explaining the quadrupole splitting on the basis of complete $5s5p^3$-hybridization with partial ionicity of the Sn–Hal bond is the more nearly correct, since direct determination of the molecular weights of these compounds has shown that, unlike SnF_4, they have a monomeric structure [74, 75]. When treated in this way, assuming that $Q(Sn^{119*}) = 8 \cdot 10^{-26}$ cm^2 [67], our experimental data on the quadrupole splitting in the Ph_3SnHal series [52] lead to the following values for the ionic fraction of the Sn–Hal bonds: x = 0.55 (I), 0.7 (Br, Cl), and 1 (F). Since the value of $x \approx 1$ is obtained for Sn–F, it may be concluded that the value of $Q(Sn^{119*})$ is in any case not too high. It would be desirable to make a direct determination of the effective charges of the halogen atoms in the Ph_3SnHal series

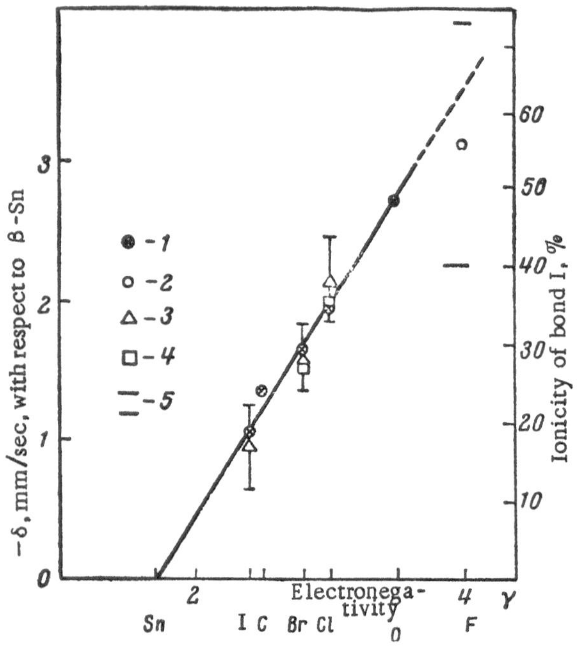

Fig. 20. Chemical shifts in SnX_4 compounds as a function of the electronegativity of X and the degree of ionicity of the Sn−X bonds. 1) Data of [52]; 2) data of [67] — on the left-hand scale, i.e., on the Mössbauer effect; 3) data of [100]; 4) data of [101] — on the right-hand scale, i.e., found by other methods; 5) position of the two components of the Mössbauer spectrum for SnF_4 [67].

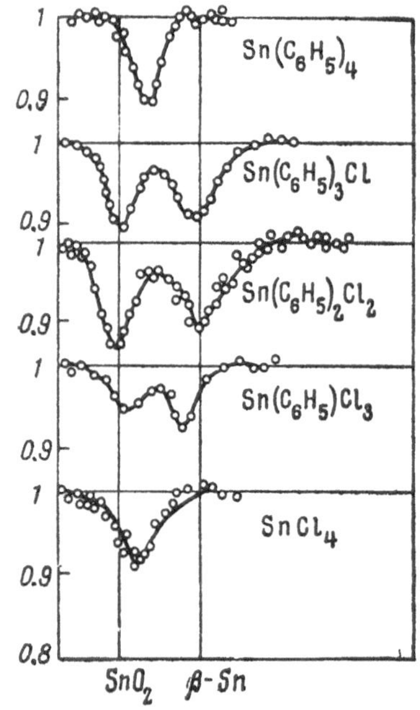

Fig. 21. Asymmetry in the doublet splitting of the Mössbauer spectrum lines in organotin compounds.

and, by comparing the results obtained with the quadrupole splitting of the Mössbauer spectra, make a direct determination of the quadrupole moment of the excited Sn^{119} nucleus. However, even now it may be concluded from the existing data on the degree of ionic character in the halogen atom bonds in various compounds and, in particular, the data given in Fig. 20 on the degree of ionicity of the Sn−Hal bonds, that $Q(Sn^{119*}) = (14 \pm 2) \cdot 10^{-26}$ cm^2. The quadrupole splitting for SnF_4 (1.7 mm per sec [67] to 2 mm/sec [124]) is due to the contribution of structures in which the Sn−F bonds are not all equivalent. This result in comparing the Mössbauer spectra of SnF_4 with other tin halides is in accordance with the idea that SnF_4 is an inorganic coordination polymer [102], in which each Sn atom is bound with six F atoms (octahedron with D_{4h} symmetry), of which two of the F atoms no longer have additional bonds, while the other four F atoms form bridge-type bonds between the tin atoms. Bridge-type structures and, in particular, those having polycentric orbits, which, according to Ya. K. Syrkin [103], are in general of exceedingly wide occurrence, may in principle be formed in a number of organotin compounds. Since, however, structure analysis has shown that a tetrahedral configuration (for example [104]) is present in the series of compounds of the type R_iSnHal_{4-i} [for example, $(CH_3)_3SnI$, $(CH_3)_3SnCl$, and $(CH_3)_2SnCl_2$], it is obviously necessary in these cases to look for an explanation of the quadrupole splitting within the framework of the four sp^3-bonds of the tin atom.* In addition, a determination of the molecular weight of Ph_3SnCl in our experiments [74, 75] showed that in this case at least we are dealing with a monomer.

* The interpretation of the Mössbauer spectra will of course be somewhat different for those organotin compounds in which the molecules do not have complete sp^3-hybridization or where there is sp^3d-hybridization such that the tin atoms have the coordination number 5. Nevertheless, in nuclear magnetic resonance (NMR) experiments, it has been shown [129, 130] from measurements of the spin−spin interaction of the protons in the methyl groups with the Sn^{117} and Sn^{119} nuclei that the degree of $5s5p^3$-hybridization in compounds of the type $(CH_3)_iSnCl_{4-i}$ changes with the number i. Further, a number of studies have been published recently [130-137], in which x-ray, IRS, and NMR data lead to the conclusion that there is a coordination number of 5 in some of the derivatives of trimethyltin, for example $(CH_3)_3SnF$, $(CH_3)_3SnOCOCH_3$, and other organotin compounds.

In concluding this section, we note that organotin compounds were the first compounds for which it was possible to interpret the asymmetry and the doublet splitting of the Mössbauer spectra in isotropic polycrystalline samples [52, 73-75]. We have observed strong asymmetry in the splitting for a number of derivatives of the type R_iSnHal_{4-i} (see, for example, Fig. 21), persisting even when the test material was ground up or dissolved and, hence could not be accounted for by the orientation of the samples. Magnetic weighing and studies of the samples by electron paramagnetic resonance have shown that the asymmetry observed could not be due to ferromagnetic or paramagnetic impurities either, as assumed for example in [122, 123]. Asymmetry of the two quadrupole splitting peaks in the spectrum of a polycrystalline isotropic sample may be produced by anisotropy of the Mössbauer effect in the single crystals themselves. As a matter of fact, as may be shown from [105], if the axis of the axially symmetric electric field of a single crystal is oriented at an angle ϑ to the direction of the γ quanta, the ratio of the π- to the σ- transition intensities is

$$\frac{i_\pi(\vartheta)}{i_\sigma(\vartheta)} = \frac{2\sqrt{5}\,\bar{P}_0(\cos\vartheta) + \bar{P}_2(\cos\vartheta)}{2\sqrt{5}\,\bar{P}_0(\cos\vartheta) - \bar{P}_2(\cos\vartheta)} = \frac{1+\cos^2\vartheta}{\frac{5}{3}-\cos^2\vartheta}, \tag{31}$$

where the $\bar{P}_L(\cos\vartheta)$ are normalized Legendre polynomials. If this ratio is averaged over the angles, which is equivalent to using an isotropic polycrystalline sample, we obtain

$$\frac{i_\pi(\text{total})}{i_\sigma(\text{total})} = \frac{\int_{-1}^{+1} i_\pi(\vartheta)\, d\cos\vartheta}{\int_{-1}^{+1} i_\sigma(\vartheta)\, d\cos\vartheta} = 1.$$

However, if the Mössbauer effect is anisotropic, i.e., the probability of γ-transition without phonon excitation itself depends on the angle ϑ, we have

$$\frac{i_\pi(\text{total})}{i_\sigma(\text{total})} = \frac{\int_{-1}^{+1} i_\pi(\vartheta)\, f(\cos\vartheta)\, d\cos\vartheta}{\int_{-1}^{+1} i_\sigma(\vartheta)\, f\cos\vartheta)\, d\cos\vartheta} = F\,[f(\cos\vartheta)] \neq 1, \tag{32}$$

i.e., asymmetry occurs in the doublet splitting. Here, neglecting fluctuations of the electric field gradients, the two peaks are of similar form [i.e., we have $y = f(x)$ for one peak and $y = \text{const} \cdot f(x)$ for othe other peak], both for uniaxial crystals and for those cases where the crystal field of the single crystal is not axially symmetric. However, if we also take into consideration fluctuations in the gradients, the forms of the two peaks are different in addition to the difference in percent absorption at the minimum. This question has been given a detailed theoretical treatment by S. V. Karyagin [73], and an experimental confirmation of the explanation given for the asymmetry has been provided [74, 75] by comparing the spectra for Ph_3SnCl in four different forms:

a. isotropic polycrystalline sample at an angle of 90° to the direction of the γ beam;
b. isotropic polycrystalline sample at an angle of 45° to the direction of the γ beam;
c. partially oriented sample at an angle of 45° to the direction of the γ beam.
d. partially oriented sample at an angle of 45° to the direction of the γ beam.

Changing the angle at which the sample is oriented to the γ beam did not produce any change in the asymmetry of the peaks on going from a to b, but did produce a change on going from c to d. On the other hand, the partial orientation of the sample produced a change in the asymmetry of the peaks at both 90° and 45°. Grinding up the sample and destroying the partial orientation produced a change in the spectra observed for c and d and restored the picture found for a, which is identical to that of b.

The fact that the results of experiments for a and b are identical eliminates any simple explanation of the asymmetry of the two peaks within the framework of quadrupole splitting due to the presence of anisotropy and hence of a definite orientation of the sample with respect to the γ beam. However, the change in the spectra on going from b to c and then to d eliminates any possibility of explaining the asymmetry in terms of trivial superposition of two singlet lines with different chemical shifts. Thus, the results presented eliminate any need of interpreting the asymmetric doublet splitting as a superposition of two different chemical shifts – an interpretation which, as we have seen at least in the case of $FeCl_3 \cdot 6H_2O$, leads to obvious misunderstandings. In addition, these results open up new possibilities of investigating the characteristics of single crystals by observing the Mössbauer spectra of polycrystalline samples. Moreover, the nature of the asymmetry in the quadrupole splitting is a new additional parameter of the Mössbauer spectra which depends on the structure of the individual molecules. It has been shown, for example by our experiments with various derivatives of diphenyltin oxide, that introducing halogen groups into benzene rings has no appreciable effect on the magnitude of the chemical shifts and the quadrupole splitting, but it shows up sometimes as a change in the type of asymmetry in the splitting.

Many other elements in addition to iron and tin have already been found to be suitable for studies of the Mössbauer effect. However, the basic problem of these studies so far has been simply to establish the existence of the effect and investigate the basic nuclear characteristics of the γ-transitions (decay scheme, level width, etc.). Only for gold Au^{197} have the changes already been studied in the chemical shifts as a function of the composition of the lattice (Au, Pt, steel, Fe, Co, Ni) into which the gold was introduced as a small impurity [106, 107]. Accordingly, in this review we are limiting ourselves to a detailed discussion of the results for iron and tin, in particular cases mentioning the other elements only in the next concluding section, which is devoted to some of the prospects for using the Mössbauer effect in chemistry.

CHAPTER VII

SOME PROSPECTS FOR USING THE MÖSSBAUER EFFECT IN CHEMISTRY

Problems in the Nature of Chemical Bonds

Since the basic parameters of the Mössbauer spectra such as the chemical shifts and the quadrupole splitting are to a considerable degree determined by the structure of the valence electron shells of the Mössbauer atoms, a first and natural possibility of applying the Mössbauer effect to chemistry is the investigation of the nature of the bonds formed by the atoms. Here, the simplest approach is to make a distinction between two types of bonds, ionic and covalent, and to find the contribution made by each type. Without special calculations, however, even this simplified approach is suitable only for explaining why different compounds have a certain relative position in a series of values of chemical shifts or quadrupole splitting.

Any attempts at quantitative interpretation require rather laborious calculations of the electron density $|\psi(0)|^2$ and the electric field gradient $q = \partial^2 V/\partial z^2$ in the region where the nucleus is located. With the appearance of rapid electronic computing machines, such calculations are facilitated, and it is necessary to make them.

In addition, it is very desirable that all types of experiments permitting an independent determination of such nuclear characteristics of Mössbauer emitters as the relative dimensional change $\Delta R/R$ and the quadrupole moment Q be performed. Thus, for example, it would be very interesting to measure the isotope shifts in the optical spectra of different compounds of Mössbauer atoms. The value of ΔR R with change in mass number may be found with good accuracy, and thus from the values of the isotope shifts it would be possible to find the values of $|\psi(0)|^2$, which could then be used to interpret the Mössbauer spectra. In finding the values of the quadrupole moments and the subsequent determination of the electric field gradients, it would be important to investigate Coulomb excitation of Mössbauer levels. As we know, the data on quadrupole splitting in conjunction with the simplified methods of calculating the gradients [56], widely used at the present time, still do not permit us to arrive at a unique conclusion as to the nature of the chemical bonds, since the reduction in the main contribution to the values of q, usually due to p_z-electrons, may be caused either by strengthening the ionic nature of the bonds or by sp-hybridization of the valence bonds. However, the Mössbauer spectra have an additional parameter, the chemical shift, which sometimes makes it possible to decide between the above two possibilities. For example, in the case of quadrivalent tin already discussed, taking off two 5p-electrons ($5s^2 5p^2 \rightarrow 5s^2$) increases the value of $|\psi(0)|^2$, while hybridization ($5s^2 5p^2 \rightarrow 5s5p^3$) acts in the opposite direction. The ability to distinguish between the effects of s-electrons (chemical shifts) and p- and d-electrons (quadrupole splitting) in the valence shells, i.e., the ability to determine whether increasing or reducing the inhomogeneity of the intramolecular electric field is accompanied by an increase or a decrease in the electron density in the region where the Mössbauer nucleus is located, is of course an advantage over the other nuclear resonance methods. Further, a number of interesting possibilities for structural chemistry may be opened up by making a systematic study of the relation between the asymmetry of the two components of the quadrupole splitting of the Mössbauer spectrum lines (or of the anisotropy in the Mössbauer effect) and the structure of the molecules in question.

In this connection it is particularly desirable to develop generalized methods of calculation which make it possible to calculate the parameters of both the chemical shifts and the quadrupole splitting of the Mössbauer spectrum lines.

In addition, it must not be forgotten that dividing the chemical bonds into ionic and covalent is in itself a rather crude simplification, since no account is taken of the possibility of forming donor–acceptor and dative bonds, nor of bonds formed by polycentric orbits, nor of other forms of the chemical bond observed in recent years [96]. Not one of the methods now in existence for investigating electron structures, including the Mössbauer effect, can give an exhaustive answer to the question of the nature of the chemical bond, and there must be comparison and mutual complementation of the information that they give, as demonstrated above in the case of the Mössbauer spectra and the effective charges of iron. The Mössbauer effect provides the simplest experimental method of observing even very small changes in the value of $|\psi(0)|^2$, which is relatively weakly dependent on the macroscopic structure of the sample, but very sensitive to screening of the Mössbauer nuclei by the electrons in the various molecular orbits. Accordingly, in addition to the studies already made on the dependence of $|\psi(0)|^2$ on the nature of the neighbors of the Mössbauer atoms, a very promising problem, mentioned in Chapter IV, is the observation of the changes in the chemical shifts and the quadrupole splitting as a function of the nature of the more remote chemical bonds. This prospect is particularly attractive for the various organotin compounds in view of the fact that tin, with the exception, it is true, of the very important role played by d-electrons, is an analog of carbon and much of the information thus obtained can be made direct use of in organic chemistry.

It would be interesting to introduce into organotin molecules various groups showing a strong inductive or mesomeric effect, or large inductomeric polarizability, and then try to find quantitatively what effect they have on the electron density at the tin nuclei. Here it is desirable to compare the effect of the groups, not merely as a function of how far they are from the tin atoms, but at the same time as a function of the nature of the intermediate bonds. It is, for example, possible to find how the electron density is transmitted along a fatty chain without disturbing a conjugated chain, and through an aromatic ring, in which the substituents are in the O–P or M position, etc.

Examples of the changes in the chemical shifts and the quadrupole splitting on going from alkyl to aryl radicals, which are the stronger electron acceptors, have already been given above (page 39).

One more example of this type was observed in [123], in which an increase in quadrupole splitting was observed in the series $(C_4H_9)_2Sn(CH_3COO)_2$; $(C_4H_9)_2Sn(CH_2ClCOO)_2$; $(C_4H_9)_2Sn(CCl_3COO)_2$. An increase of this sort in the quadrupole splitting (see Fig. 19) is a natural consequence of the increase in ionic character of two out of the four valence bonds of the tin. There is a particularly large increase in the quadrupole splitting on going from organotin derivatives of carboxylic acids to the organotin salts $(C_4H_9)_2SnSO_3$ and $(C_4H_9)_2SnSO_4$ [108].

In all cases, the tin atom is here bound with two alkyl radicals (butyl) and two oxygen atoms, but the ionic character of the bond with strong inorganic acid residues is naturally much greater. It is interesting, however, that introducing chlorine into an acetic acid residue, and, to an even greater extent, going to dibutyltin sulphate $(C_4H_9)_2SnSO_4$, i.e., going to stronger and stronger electron acceptor groups, leads, according to [108, 123], to an increase in electron density in the vicinity of the tin nucleus. This question requires further experimental refinements. It is not impossible that the Mössbauer effect in Sn^{119} will turn out to be too "crude" an instrument ($\Gamma/E_r = 10^{-12}$) for making very thorough observations on the effect of remote chemical bonds in any given molecule, but there is no doubt that there is a general possibility in principle of detecting the changes which they produce in the Mössbauer spectra.

As we have mentioned above, the principal application of the Mössbauer effect in chemistry is, apparently, the study of metalloorganic and complex compounds. In the field of metalloorganic compounds there is substantial interest in making a comparison between the general nature of the metallocarbon bonds, since there is a great difference between them for transition metals and metals belonging to the basic groups.

Recently, A. N. Nesmeyanov [109] gave a very complete review of the state of this problem, on the basis of which we can easily lay out the main lines of application of the Mössbauer effect. Thus, the Mössbauer effect could be used to compare the acetylenide complexes of transition metals of the type $K_4Fe(C\equiv CH)_6$ (similar to the cyanide complexes, and formed apparently with the d-levels of Fe) with the alkenyl compounds of metals of the type $Sn\text{-}(CH=CHCl)_4$, where there is a σ-bond resulting from sp^3-hybridization. It would also be interesting to make a systematic comparison of the Mössbauer spectra of cyclopentadienylides of metals, $Me(C_5H_5)_2$, with various types of ferrocene-like "sandwich" structures, of the σ-derivatives of the type of dicyclopentadienyl tin, and of the "sandwich," but ionic structures of the type $Me^{\cdot\cdot}[(C_5H_5)^-]_2$. An interesting feature, already observed by means of the Mössbauer effect and now being carefully studied, is the great similarity between the Mössbauer spectra for ferrocene and the phthalocyanine of bivalent iron.

It has been found that iron carbonyl, which has very nearly the same quadrupole splitting as ferrocene, shows a quite different chemical shift – the value of $|\psi(0)|^2$ for carbonyl is much higher (see Fig. 15).

In order to compare the iron bonds in π-cyclopentadienyl compounds and those in carbonyls, it is desirable to investigate the Mössbauer spectra of mixed cyclopentadienylide carbonyls, polycarbonyls, and polyferrocenes. Related to carbonyls (and complex acetylenides) are such complex compounds as cyanide salts. There are even mixed carbonyl-cyanide derivatives with anions of the type $[Fe(CO)(CN)_5]^{3-}$. Iron carbonyl, which has nearly the same chemical shift as the cyanide complexes, is nevertheless different, having a quadrupole splitting many times larger. As we have seen above in the case of ferrocene, the quadrupole splitting of π-cyclopentadienylides yields easily to a semiquantitative treatment based on the calculations of [98, 99]. It is accordingly a very good idea to investigate the changes in quadrupole splitting during a gradual transition from the carbonyl to the cyanide complexes, passing through the mixed forms, and passing finally to heterometallic carbonyls and carbonyls that are partially substituted by nitrogenous bases, etc. Among the metalloorganic compounds of nontransition elements, it would be interesting to continue the above comparison of chemical shifts and quadrupole splittings in the series SnR_iX_{4-i}, where R = aryl, alkyl, or vinyl. The Me–C bond strengths are considerably less in alkyl derivatives, i.e., the nature of the bond is not completely determined by the difference in electronegativity between the metal and the carbon, but depends to a large extent on the nature of the organic radical. Aryl and vinyl radicals are more easily removed in the form of anions than alkyl radicals, but it is more difficult to remove them in the form of radicals or cations. As a comparison standard, Mössbauer spectra should be taken of a hybrid of tin and mixed H_iSnR_{4-i} derivatives.

The Mössbauer effect in iron is of particular interest since iron, in one form or another, forms part of very many biologically important structures. It is true that the amount of iron in these structures is very small, so that for the investigations to succeed it would be necessary to synthesize compounds enriched in Fe^{57}, but even now the first results have been obtained in this field by using hemin. One of the problems which use of the Mössbauer effect can greatly aid in solving is determining whether or not, as shown by L. A Blyumenfel'd's data [110, 111], there is anything specific about the structure of the electron shells or the origin of the local magnetic fields at the iron nuclei occurring as small impurities in DNA and the DNA complexes formed with proteins.

In speaking of the problems of chemical structure, we must again mention the polymeric coordination compounds, based not on ordinary covalent bonds but on donor–acceptor, two-center, two-electron bonds, as well as the various polycentric bonds including, in particular, electron deficient polymers. The bridge structures produced by these bonds are of such wide occurrence that if the gross formulas of inorganic compounds represent their true structure, it must not be regarded as the rule but rather as a rare exception. For this reason, for example, Ya. K. Syrkin [103] states that "the concept of trivalent iron is evidently out of date, since iron does not form three bonds." Thus, according to [103], in $FeCl_3$, as a result of covalent and donor–acceptor bonds, each iron atom is bound with six chlorine atoms, while each chlorine atom forms bridging bonds between two or three iron atoms. As was shown above for $SnHal_4$, the Mössbauer effect enables us to draw a fairly sharp dividing line between bonds of the usual, two-electron type (Hal = Cl, Br, I) and the formation of coordination structures (SnF_4). It is not impossible that a similar difference exists between $FeCl_3$, with $\Delta = 0$, and $FeCl_3 \cdot 6H_2O$, for which, apparently, $\Delta = 0.95$ mm/sec. Accordingly, a systematic comparison must be made between the Mössbauer spectra and the chemical shifts and quadrupole splitting calculated under various assumptions as to the nature of the bonds. Compounds should also be studied for which it is possible to have coordination polymer structures, for example, the tin alkoxydes $Sn(OR)_4$, in which it is to be expected that the Mössbauer spectrum will be similar to SnF_4. A number of coordination polymers may be formed in solutions such, for example, as the mono-, di-, and trimeric complexes, $SnOH^+$, $Sn_2OH_4^{2+}$, and $Sn_3(OH)_4^{2+}$, which occur in the hydrolysis of $Sn^{\cdot\cdot}$, and the complexes $FeOH^{2+}$, $Fe(OH)_2^+$, and $Fe_2(OH)_2^{4+}$ occurring in the hydrolysis of $Fe^{\cdot\cdot\cdot}$ [102]. The formation of these complexes will also affect the Mössbauer spectra, and in this connection we immediately encounter possibilities of investigating both structural and kinetic problems (the properties of the intermediate products), to which we shall later turn. For the present, however, we shall point out the particular interest for the chemistry of complex compounds in observing and making use of the Mössbauer effect in platinum. We shall give one example here of a quite well-known coordination polymer, trimethylchlorplatinum, $Pt(CH_3)_3Cl$. It is analogous in structure to tetramethylplatinum, where polycentric orbits are used to form the tetrameric molecule shown in Fig. 22a. A similar tetramer may be prepared from trimethylchlorplatinum. X-ray analysis has shown that the chlorine and platinum atoms are located at the vertices of the cube in this compound. However, as may be seen from Fig. 22, this tetramer can in principle exist in two forms, both in the form $[Pt(CH_3)_3Cl]_4$ (Fig. 22b), in which the bridging bonds are formed by chlorine atoms and each platinum atom is bound directly to three chlorine atoms and three methyl radicals, and in the form $\{Pt[(CH_3)_2Cl]CH_3\}_4$ (Fig. 22c), in which the bridging bonds are formed by methyl radicals and each platinum atom is bound to one chlorine atom and five methyl radicals. This peculiar bridging isomerism would correspond to two different chemical shifts in the Mössbauer spectra of platinum. Thus, the Mössbauer effect could, in this case, help to observe the existence of new chemical phenomena.

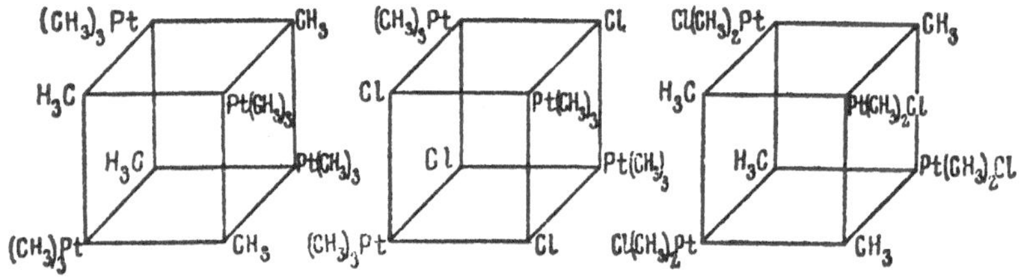

Fig. 22. Bridge stereoisomerism possible for the trimethylchlorplatinum tetramer. a) Structure of $[Pt(CH_3)_4]_4$; b, c) two possible forms for the structure of $[Pt(CH_3)_3Cl]_4$.

Problems in Chemical Kinetics, Radiation Chemistry, and the Physical Chemistry of Polymers

In addition to problems in the structure of chemical compounds, the Mössbauer effect can, beyond doubt, also be used in chemical kinetics and radiation chemistry. In addition to the possibility of taking kinetic curves directly, all in one experiment (from the frequency of the counts at some fixed characteristic rate of motion), there are particularly interesting prospects here of observing unstable intermediate products. When reactions are being carried out in the liquid phase, we have to stop the process and freeze the mixture for each observation of the Mössbauer spectrum. However, with topochemical processes (in particular, and what is especially interesting, with radiation topochemical processes), the changes in the Mössbauer spectrum may be observed continuously during the reaction. An example of the possibility is provided by our observation of a sudden change in the form of the Mössbauer spectra while a solid equimolecular mixture of $SnPh_4$ + SnI_4 was being irradiated with 1.6-MeV electrons [52]. The change came from the formation of various disproportionation reaction products of the type Ph_iSnI_{4-i}. Radiation chemical transformation of dibutyltin sulphate $(C_4H_9)_2SnSO_4$ and of a tin-containing polymer based on methylmethacrylate was detected by observing the Mössbauer spectra in the paper by L. S. Polak, V. S. Shpinel' and co-workers [122].

Among other radiation chemical reactions which may be investigated with the aid of the Mössbauer effect, we can mention the very peculiar radiation synthesis of organotin compounds performed by K. A. Kocheskhov, et al. [112] by irradiating tin in halogen-substituted hydrocarbon derivatives:

$$Sn + 2R\,Hal \rightarrow R_2Sn\,Hal_2,$$

as well as the radiation oxidation of bivalent tin by trivalent iron salts in aqueous sulfuric acid solutions, during which unstable intermediate products are assumed to be formed — derivatives of mono- and trivalent tin [113]. It is also possible in various solvents to observe heterolytic exchange of the type $SnR_4 + KCl \rightarrow R_3SnCl + KR$, in which the radical plays the role of an anion, and homolytic exchange of the type Me + Me'R, in which the radical goes from one metal to another. Formation of a polymer will not primarily affect the chemical shifts or the quadrupole splitting of the Mössbauer spectrum lines but will affect the recoilless resonance gamma fluorescence probability itself, i.e., it affects the Debye–Waller factor. In this connection, it would also be desirable to compare the probability of the Mössbauer effect for different molecular weight fractions of polymer samples.

It would be extremely interesting to use the Mössbauer effect to observe polymerization (for example, of monomers containing tin) in the solid phase, since in experiments on radiation solid phase polymerization we usually do not know whether the polymer is formed during the actual irradiation or during the subsequent thawing, for example, at the phase transition points.

Substantial interest is presented by the arylation of $SnCl_2$ in organic solvents by salts of iodonium type, investigated by O. A. Reutov and co-workers [114],

$$2Ar_2ICl + 2SnCl_2 \rightarrow Ar_2SnCl_2 + 2ArI + SnCl_4,$$

which the authors assumed to proceed with the formation of trivalent tin derivatives as an intermediate product. The Mössbauer effect could give much new information on the kinetics and mechanism of these reactions.

Use of the Mössbauer effect in polymerization studies is also promising since wider and wider use is being made of bimetalloorganic complexes containing tin, of the type of the heterogeneous catalyst $Sn(C_4H_9)_4 + TiCl_4 + AlCl_3$ [115] or of the homogeneous catalysts $Sn(C_6H_5)_4 + (C_6H_5)_2VCl_2 + AlBr_3$ [116] and $Sn(C_6H_5)_4 + VCl_4 + AlHal_3$ [117], as efficient polymerization catalysts. Tetrabutyl tin has no appreciable ability to reduce $TiCl_4$, but it is a good alkyl donor, and when it reacts with the electrophilic $AlCl_3$ it forms a high-activity catalyst. The state of the catalytic complexes, and of the change in electron structure of the tin atoms during the actual polymerization, will evidently be somewhat clarified when Mössbauer spectra are taken.

Experiments made on polymer molecules containing Mössbauer atoms (for example, tin) are of special interest. It is possible here, in particular, to observe how the line shift (from the recoil energy of the macromolecule) and, at the same time, the Doppler broadening vary with molecular weight of the molecules or "cross linking" of the polymer. "Doctoring up" the line broadening in this way to $D \approx 0.01\text{-}0.1\ k\theta$ (where θ is the Debye temperature of the emitter) would make possible – in conjunction with measurements at velocities of the order of $(D/E)c$ – a direct study of the phonon spectrum of the emitter. It would also be possible to find out whether the macromolecule as a whole experiences recoil or whether deformational vibrations occur (which have an activation energy and are thus suppressed at low temperatures), and at the same time to compare, from this point of view, polymers containing tin in the main chain and in the side groups.

The anisotropy of the Mössbauer effect and the accompanying asymmetry in the quadrupole splitting make possible a direct study of the special structural features of the internal electric fields in the molecules of oriented polymers.

It is not impossible in principle to observe the Mössbauer effect in a liquid, since long-wave excitations are propagated in a liquid just as in a solid, and the only "dangers" to the effect are the short-wave excitations, which propagate in a diffusional way, i.e., with an activation energy, and are thus suppressed at low temperatures. It would accordingly be very interesting to investigate macromolecules dissolved in a liquid at very low temperatures, e.g., a polymer containing tin in liquid methane, as well as experiments with viscous liquids.

Thus, in investigating the Mössbauer effect in polymer molecules, we can certainly obtain some very interesting information on how the recoil energy is transmitted in such molecules themselves and to collective forms of excitation in the surroundings. Note further the possibility, mentioned above in connection with [69], of using the Mössbauer effect to investigate the various chemical consequences of nuclear transformations, recently, in particular, realized in [118]. It would be interesting to see whether or not any marked changes in the spectra occur when the Mössbauer atoms are in a zone adjacent to the tracks of fission fragments or of the strongly ionized particles produced when thermal neutrons are captured by Li^6 and B^{10} nuclei, since in this zone (with a diameter up to about 100 A) there is short-time (10^{-10} sec) strong heating followed by rapid quenching [119].

Exceedingly interesting possibilities are opened up by a study, no longer of the chemical shifts or the quadrupole splitting, but of the actual value of f, i.e., the probability of the Mössbauer effect in nuclei included as impurities in arbitrary matrices of heavy or light nuclei. Here we can even think of infinitesimally small concentrations of those Mössbauer impurity nuclei which are most easily realized experimentally, the nuclei added being not absorbing but emitting γ-active nuclei.

It has been shown by Yu. M. Kagan and Ya. A. Iosilevskii [120, 121] that an appreciable Mössbauer effect exists even in the case when the matrix enclosing the Mössbauer radiators or absorbers consists entirely of light nuclei. It turns out that the value of W (upon which the recoilless resonance fluorescence probability is exponentially dependent) in the Debye–Waller factor increases with increase in mass m of the nuclei of the light matrix, not as $W \sim 1/m$ [as might be expected from the proportionality $W \sim R$; see (20)], but more weakly – as $W \sim 1/\sqrt{m}$. The theory [120, 121] proved to be in excellent agreement with the results of the experiments already mentioned on small amounts of gold added as impurities [106, 107], as well as of the experiments with iron in beryllium [90].

Recently, careful measurements of the probability of the Mössbauer effect were conducted over a wide temperature range, both for the case of strong emitters (Sn^{119} in vanadium [138]) and for the case of weak emitters (Sn^{119} in gold, platinum, thalium [139] and Fe^{57} in gold [140]). Good agreement between the experimental data and theoretical results [120, 121] was obtained in all cases. The work in [138] should be particularly noted, since the phonon spectrum for vanadium is well-known from studies of slow-neutron scattering, and thus the comparison of theory with experiment could be carried out thoroughly. We note that the results of these works evidently confirm the hypothesis that in metallic alloys which form solid solutions with small concentrations of one of the components, power constants change very little.

It is thus, in principle, possible to use the Mössbauer effect to investigate the properties of even the lightest lattices. The change in the probability of the Mössbauer effect as a function of temperature, pressure, and other external conditions serves in this case as a detector of the various phase transitions in the matrix and gives the characteristics of the long-range internal fields (for example, the magnetic fields acting on the diamagnetic Mössbauer atoms in a ferro- or antiferromagnetic matrix) and of the other macroscopic properties of various matrices. In the field of the physical chemistry of surface phenomena, there may be substantial interest in a study of the probability and anisotropy of the Mössbauer effect when using "two-dimensional" emitters or absorbers – mono- and polymolecular surface layers.

Finally, attention must be called to the highly probable analytical applications of the Mössbauer effect in chemistry, for example, in those cases where all possible types of disproportionation reactions (of the type $SnR_4 + SnX_4 \rightarrow R_iSnX_{4-i}$) occur forming products in which the Mössbauer spectra differ substantially from the original materials.

The number of examples given could easily be increased, but probably enough have already been given to give some idea of the rich possibilities that exist for using the Mössbauer effect – this discovery of nuclear physics, so remarkable for its simplicity and ingenuity – in chemistry.

APPENDIX

Elements for Which the Mössbauer Effect Has Been or Should Be Observed

1 H										2 He
3 Li	4 Be	5 B	6 C	7 N	8 O	9 F				10 Ne
11 Na	12 Mg	13 Al	14 Si	15 P	16 S	17 Cl				18 Ar
19 K	20 Ca	Sc 21	Ti 22	V 23	Cr 24	Mn 25	~~Fe 26~~	Co 27	~~Ni 28~~	
Cu 29	~~Zn 30~~	31 Ga	~~32 Ge~~	33 As	34 Se	35 Br				~~36 Kr~~
~~37 Rb~~	38 Sr	Y 39	Zr 40	Nb 41	Mo 42	Tc 43	~~Ru 44~~	Rh 45	Pd 46	
Ag 47	Cd 48	49 In	~~50 Sn~~	~~51 Sb~~	~~52 Te~~	~~53 I~~				~~54 Xe~~
~~55 Cs~~	56 Ba	* ~~La 57~~	~~Hf 72~~	~~Ta 73~~	~~W 74~~	~~Re 75~~	~~Os 76~~	~~Ir 77~~	~~Pt 78~~	
~~Au 79~~	~~Hg 80~~	~~81 Tl~~	82 Pb	83 Bi	84 Po	85 At				86 Rn
87 Fr	~~88 Ra~~	** Ac 89								

* 58-71 Lanthanides
** 90-103 Actinides
} Mössbauer effect should be observed in almost all these elements

[box with diagonal line] Mössbauer effect should be or has already been observed.

Isotopes for Which the Mössbauer Effect Has Been or Should Be Observed

Nucleus absorbing γ quanta	% content	E_r, keV	$T_{1/2}$ of level, sec	Γ/E_r	Conversion coefficient α	$R \cdot 10^{-2}$, eV	$\sigma_0/(1+\alpha) \cdot 10^{19}$, cm^2	Mother nucleus	$T_{1/2}$ of mother nucleus
Fe^{57}	2.17	14.4	1 (−7)	3.1 (−13)	15	0.19	15	Co^{57}	270 days
Ni^{61}	1.25	71	5.2 (−9)	1.2 (−12)	K : 0.11	4.4	6.6	Cu^{61}	3.3 hr
Zn^{67}	4.11	93	13.4 (−6)	5.2 (−16)	K : 0.63	6.9	1.2	Ga^{67}	78 hr
Ge^{73}	7.67	13.5	3.1 (−6)	7.1 (−15)	3600	0.13	0.03	As^{73}	76 days
Kr^{83}	11.55	9.3	1.5 (−7)	4.7 (−13)	10	0.055	21.0	Rb^{83}	100 days
	11.55	9.3	<1 (−7)		10	0.055	21.0	Rb^{83}	100 days
Rb^{85}	72.1	150				14	0.73	Sr^{85}	65 days
Ru^{99}	12.7	90	2 (−8)	3.7 (−13)		4.3	1.0	Rh^{99}	16 days
	12.7	89				4.3	1.0	Rh^{99}	15 days
Ru^{101}	17.0	127	1.4 (−9)	2.6 (−22)	K : 0.4	8.5	0.97	Rh^{101}	4.3 days
Ag^{107}	51.35	93	44.3	1.1 (−22)	16	4.3	0.67	Ag^{107m}	44 sec
Ag^{109}	48.65	88	39.2	1.3 (−22)	14	3.8	0.85	Ag^{109m}	40 sec
Sn^{117}	7.57	161			K : 0.13	12.0	1.7	Sn^{117m}	14 days
Sn^{119}	8.58	24	1.9 (−8)	1.0 (−12)	7.3	0.26	10.0	Sn^{119m}	250 days
Sb^{123}	42.75	161				11.0	0.71	Sn^{123}	136 days
Te^{123}	0.87	159	1.9 (−10)	1.5 (−11)	K : 0.17	11.0	1.7	Te^{123m}	104 days
Te^{125}	6.99	35	1.6 (−9)	8.3 (−12)	K : 12	0.52	3.1	Te^{125m}	58 days
Te^{127}	100	59			1.9	1.5	3.2	Te^{127m}	105 days
I^{129}	Radio-active ($1.6 \cdot 10^7$ years)	26.8	2.7 (−8)	9 (−13)	5	0.3	4.3	Te^{129m}	70 min
Xe^{129}	26.44	40	7 (−10)	2.4 (−11)		0.67	30.0	Xe^{129m}	8 days
								I^{129}	$1.6 \cdot 10^7$ years
	26.44	40				0.67	30.0	Xe^{129m}	8 days
Xe^{131}	21.18	80	4.8 (−10)	1.2 (−11)	K : 1.73	2.6	0.70	Xe^{131m}	12 days
Cs^{133}	100	81	6.0 (−9)	9 (−13)	K : 1.5	2.7	1.1	Ba^{133}	9.5 years

(continued)

Nucleus absorbing γ quanta	% content	E_r, keV	$T_{1/2}$ of level, sec	Γ/E_r	Conversion coefficient α	$R \cdot 10^{-2}$, eV	$\sigma_0/(1+\alpha) \cdot 10^{19}$, cm^2	Mother nucleus	$T_{1/2}$ of mother nucleus
La^{139}	99.9	163	1.5 (−9)	1.9 (−12)	K : 0.22	10.0	0.56	Ce^{139}	140 days
Pr^{141}	100	145			0.37	8	1.6	Ce^{141}	33 days
Nd^{145}	8.29	67	3.3 (−8)	2.1 (−13)	K : 3.3	1.7	0.63	Pm^{145}	18 years
		72	<1 (−9)		K : 3.3	1.9	0.82	Pm^{145}	18 years
Sm^{149}	13.84	22	2 (−8)	1.5 (−12)	12	0.18	3.9	Eu^{149}	90 days
Sm^{152}	26.63	122	1.4 (−9)	2.7 (−12)	K : 0.7	5.3	4.8	Eu^{152}	9 hr
Eu^{151}	47.77	22			L : 12	0.17	5.2	Gd^{151}	150 days
Eu^{153}	52.23	84				2.5	4.6 ⎤	Sm^{153} ⎤	47 hr
		97	<1 (−9)		K : 0.3	3.3	2.0 ⎥	⎥	
		103	3.4 (−9)	1.3 (−12)	K : 1.2	3.7	0.7 ⎦	Gd^{153} ⎦	230 days
Gd^{154}	2.15	123	1.2 (−9)	3.1 (−12)	1.5	5.3	3.2	Eu^{154}	16 years
Gd^{155}	14.7	60				1.2	10.0 ⎤		
		87			K : 0.4	2.6	2.3 ⎥	Eu^{155}	1.7 years
		105				3.8	3.3 ⎦	Tb^{155}	5.6 days
Gd^{156}	20.47	89	2 (−9)	2.6 (−12)	K : 1.0	2.7	7.7	Eu^{156}	15.4 days
Gd^{160}	21.9	75				1.9	22.0	Tb^{160}	72 days
Tb^{159}	100	58	3.5 (−11)	2.2 (−10)	K : 6	1.1	1.5	Dy^{159}	144 days
		137	5.4 (−11)	6.2 (−11)		6.3	2.6		
Dy^{160}	2.29	87	1.8 (−9)	2.9 (−12)	K : 1.5	2.6	6.4	Tb^{160}	73 days
Dy^{161}	18.88	25.7	2.8 (−8)	6.2 (−13)		0.22	37.0	Tb^{161}	6.8 days
		74.5	3 (−9)	2 (−12)		1.8	2.9		
Ho^{165}	100	95	3.3 (−11)	1.5 (−10)	K : 1.77	2.9	1.2	Er^{165}	10 hr

(continued)

Nucleus absorbing γ quanta	% content	E_r, keV	$T_{1/2}$ of level, sec	Γ/E_r	Conversion coefficient α	$R \cdot 10^{-2}$, eV	$\sigma_0/(1+\alpha) \cdot 10^{19}$, cm^2	Mother nucleus	$T_{1/2}$ of mother nucleus
Er^{166}	33.4	80	1.8 (−9)	3.1 (−12)	K : 1.7	2.1	7.1	Ho^{166}	27 hr
Er^{168}	27.07	79.8	1.84 (−9)	3.1 (−12)	K : 2.1	2.0	6.2	Tm^{168}	9.6 days
Tm^{169}	100	8.4	4 (−9)	1.4 (−11)		0.022	700]	Yb^{169}	32 days
		118	5 (−11)	7.7 (−11)	K : 0.7	4.4	3.1]	Er^{169}	9 days
Yb^{170}	3.03	84.2	1.57 (−9)	3.5 (−12)	K : 1.6	2.2	6.6	Tm^{170}	129 days
Yb^{171}	14.31	66.7	<5 (−7)			1.4	11.0	Tm^{171}	680 days
Yb^{172}	21.82	78.7				1.9	20.0	Lu^{172}	6.7 days
Yb^{173}	16.13	78.7				1.9	5.2	Lu^{173}	1.4 years
Yb^{174}	31.84	76.5				1.8	21.0	Lu^{174}	165 days
Lu^{175}	94.4	113.8	8 (−11)	5 (−11)	K : 1.6	4.0	0.90	Hf^{175}	70 days
Hf^{176}	5.21	88.3	1.35 (−9)	3.8 (−12)	K : 1.32	2.4	6.7	Ta^{176}	8 hr
Hf^{177}	18.5	113	4.2 (−10)	1 (−11)	K : 0.75	3.9	1.4	Lu^{177}	6.75 days
Hf^{180}	35.22	93	1.4 (−9)	3.5 (−12)	KL : 4.0	2.6	2.8	Ta^{180m}	8.15 hr
Ta^{181}	100	6.25	6.8 (−6)	1.1 (−14)	44	0.012	17	W^{181}	145 days
		136.1	5.7 (−11)	5.9 (−11)	K : 1.5	5.5	0.66		
W^{180}	0.135	102			5	3.1	2.0	Ta^{180m}	8.15 hr
W^{182}	26.4	100	1.3 (−9)	3.5 (−12)	4.5	2.9	2.2	Ta^{182}	115 days
W^{183}	14.4	46.5			9	0.63	2.3	Ta^{183}	5 days
		99.1			3.5	2.9	1.7		
W^{184}	30.6	111	1.3 (−9)	3.1 (−12)		3.6	9.9	Re^{184}	50 days
W^{186}	28.4	123	1.0 (−9)	3.7 (−12)	K : 0.45	4.4	5.6	Re^{186}	89 hr

(continued)

Nucleus absorbing γ quanta	% content	E_r, keV	$T_{1/2}$ of level, sec	Γ/E_r	Conversion coefficient α	$R \cdot 10^{-2}$, eV	$\sigma_0/(1+\alpha)\cdot 10^{19}$, cm^2	Mother nucleus	$T_{1/2}$ of mother nucleus
Re^{185}	37.07	125			K : 2.4	4.5	0.61	Os^{185}	94 days
Re^{187}	62.93	134	2(−9)	1.7 (−12)	K : 2.1	5.2	0.58	W^{187}	24 hr
Os^{186}	1.59	137	5.1 (−10)	6.7 (−12)	K : 0.45	5.4	4.5	Re^{186}	89 hr
Os^{187}	1.64	135				5.4	1.2	Ir^{187}	13 hr
Os^{188}	13.3	155	6.2 (−10)	4.8 (−12)	K : 0.40	6.8	3.6	Ir^{188}	41 hr
Os^{189}	16.1	135				5.4	1.4	Ir^{189}	11 days
Os^{190}	26.4	187	3.5 (−10)	7.2 (−12)	K : 0.2	9.9	2.9	Ir^{190}	11 days
Os^{192}	41.0	206	2.8 (−10)	7.7 (−12)	K : 0.16	12.0	2.5	Ir^{192}	74 days
Ir^{191}	38.5	82.6	3.9 (−9)	1.4 (−12)		1.9	1.8	Os^{191}	16 days
		129	1.4 (−10)	2.5 (−11)	K : 2.9	4.7	0.56	Pt^{191}	3 days
Ir^{193}	61.5	73	5.7 (−9)	1.1 (−12)		1.5	2.3	Pt^{193m}	3 days
		139	2.6 (−10)	1.3 (−11)	K : 2.2	5.4	0.59	Os^{193}	30 hr
Pt^{195}	33.3	99			9.0	2.7	0.50	Au^{195}	185 days
		129	5.8 (−10)	6.2 (−12)		4.6	4.4	Ir^{195}	2.3 hr
Au^{197}	100	77	1.9 (−9)	3.1 (−12)	2.5	1.6	0.59	Pt^{197}	18 hr
								Hg^{197}	65 hr
Hg^{199}	16.85	158	2.4 (−9)	1.2 (−12)	K : 0.2	6.7	2.4	Au^{199}	3.1 days
Hg^{201}	13.22	32.1				0.27	24.0	Tl^{201}	72 hr
		167.6				7.5	0.44		
Tl^{203}	29.50	279	2.9 (−10)	8.1 (−12)		21	0.63	Hg^{203}	47 days
Tl^{205}	70.50	205	1.3 (−9)	2.5 (−12)		11	1.2	Pb^{205}	$5\cdot10^7$ years
Ra^{226}		68	6.3 (−10)	1.5 (−11)		1.1	27	Ac^{226}	29 hr
								Th^{230}	$8\cdot10^4$ years

(continued)

Nucleus absorbing γ quanta	% content	E_r, keV	$T_{1/2}$ of level, sec	Γ/E_r	Conversion coefficient α	$R \cdot 10^{-2}$, eV	$\sigma_0/(1+\alpha) \cdot 10^{19}$, cm^2	Mother nucleus	$T_{1/2}$ of mother nucleus
Th^{229}		43				0.43	22	U^{233}	$1.6 \cdot 10^5$ years
		99				2.3	3.3		
Th^{232}		50				0.58	50	U^{236}	$2.4 \cdot 10^7$ years
Pa^{233}		87	3.7 (−8)	2 (−13)		1.7	4.9	Np^{237}	$2.2 \cdot 10^6$ years
Np^{237}		60	6.3 (−8)	1.8 (−13)		0.8	11]	U^{237}	6.8 days
							]	Am^{241}	470 years
U^{238}		45				0.46	62	Pu^{242}	$3.8 \cdot 10^5$ years
Pu^{239}		57				0.73	23]	Np^{239}	2.3 days
		286	1.1 (−9)	2.1 (−12)		18.4	0.9]	Cm^{243}	35 years
Am^{241}		42				0.4	14	Bk^{245}	5 days
		206				9.5	0.6		
Cm^{245}		252				14	0.4	Bk^{245}	5 days
								Cf^{249}	360 years
Cm^{246}		43				0.40	66	Cf^{250}	11 years

Note: This list, just as the list of the elements (see the previous table), is by no means fixed once and for all; it will undoubtedly be added to as the nuclear levels are investigated, although the additions will scarcely extend to elements lighter than iron. The figures in parentheses in the fourth and fifth columns give the order of magnitude of the quantity, e.g., (−7) means the factor 10^{-7}. The conversion coefficients α are either given averaged (with no letter), or for the K- or L-shells (as shown by the letter). If no value of α is given, use the value of σ_0 in the eighth column.

LITERATURE CITED

1. G. Breit, Rev. Mod. Phys. 30, 507, 1958.
2. A. R. Striganov and Yu. P. Dontsov, Uspekhi. Fiz. Nauk 55, 1955.
3. R. Weiner, Phys. Rev. 114, 256, 1959.
4. A. S. Melissinos and S. P. Davis, Phys. Rev. 115, 130, 1959.
5. C. H. Townes and A. L. Shawlow, Microwave Spectroscopy, Mc Graw-Hill, New York, 1955.
6. K. H. Hausser, Angew. Chem. 68, 728, 1956.
7. N. M. Aleksandrov and F. I. Skripov, Uspekhi Fiz. Nauk 75, 585, 1961.
8. W. J. Orville-Thomas, Quart. Rev. 11, 162, 1957.
9. V. Grechishkin, Uspekhi Fiz. Nauk 69, 189, 1959.
10. G. K. Semin and É. I. Fedin, Zhur. Strukt. Khim. 1, 252, 1960.
11. G. K. Semin and É. I. Fedin, Zhur. Strukt. Khim. 1, 464, 1960.
12. E. Segrè, Phys. Rev. 71, 274, 1947.
13. E. Segrè and C. Wiegand, Phys. Rev. 75, 39, 1949.
14. C. A. Accardo, Phys. Rev. Lett. 1, 180, 1958.
15. K. Bainbridge, M. Goldhaber, and E. Wilson, Phys. Rev. 84, 1260, 1951.
16. J. C. Slater, Phys. Rev. 84, 1261, 1951.
17. R. L. Mössbauer, Z. Physik. 151, 124, 1958.
18. R. L. Mössbauer, Naturwissenschaften 45, 538, 1958.
19. R. L. Mössbauer, Z. Naturforsch. 14a, 211, 1959.
20. R. L. Mössbauer, Uspekhi Fiz. Nauk 72, 658, 1960.
21. G. N. Belozerskii and Yu. A. Nemilov, Uspekhi Fiz. Nauk 72, 433, 1960.
22. F. L. Shapiro, Uspekhi Fiz. Nauk 72, 685, 1960.
23. R. V. Pound, Uspekhi Fiz. Nauk 72, 673, 1960.
24. V. A. Lyubimov, in: Gamma-Rays, Izd-vo Akad. Nauk SSSR, 1961, p. 682.
25. Yu. M. Kagan, in: The Mössbauer Effect [Russian translation], IL, 1962, p. 5.
26. H. Frauenfelder, in: The Mössbauer Effect, W. Benjamin, Inc., N. Y., 1962, p. 1.
27. A. J. F. Boyle and H. E. Hall, Repts. Progr. in Phys., 1962.
28. W. E. Burcham. Sci. Progr. 48, 630, 1960.
29. E. Cotton, J. phys. radium 21, 265, 1960.
30. P. P. Craig, Proceedings of the VII International Conference on Low-Temperature Physics, University of Toronto Press, Toronto, 1961, p. 22.
31. S. de Benedetti, Sci. American 202, 72, 1960.
32. W. E. Kock, Science 131, 1588, 1960.
33. I. Y. Krause and G. Lüders, Naturwissenschaften 47, 532, 1960.
34. H. Lustig, Am. J. Phys. 29, 1, 1961.
35. P. B. Moon, Nature 185, 427, 1960.
36. R. L. Mössbauer, Ann. Rev. Nuclear Sci. 12, 123, 1962.
37. G. K. Wertheim, Nucleonics 19, No. 1, 52, 1961.
38. V. F. Weisskopf, Lectures in Theoretical Physics, III, ed. W. E. Brittin and B. W. Downs, Interscience, N. Y., 1961, p. 70.

39. A. C. G. Mitchell and M. W. Zemansky, Resonance Radiation and Excited Atoms, Cambridge, 1934.
40. W. Kuhn, Phil. Mag. 8, 625, 1929.
41. G. Breit and E. Wigner, Phys. Rev. 49, 519, 1936.
42. I. Ya. Barit and M. I. Podgoretskii, Doklady Akad. Nauk SSSR 54, 591, 1946.
43. B. S. Dzhelepov, Uspekhi Fiz. Nauk 62, 4, 1957.
44. M. Goldhaber, L. Grodzins, and M. Sunyar, Phys. Rev. 109, 1015, 1958.
45. W. E. Lamb, Phys. Rev. 55, 190, 1939.
46. R. V. Pound and G. A. Rebka, Phys. Rev. Lett. 4, 337, 1960.
47. A. R. Bodmer, Nuclear Phys. 21, 347, 1961.
48. R. V. Pound and G. A. Rebka, Phys. Rev. Lett. 4, 274, 1960.
49. B. D. Josephson, Phys. Rev. Lett. 4, 341, 1960.
50. O. C. Kistner and A. W. Sunyar, Phys. Rev. Lett. 4, 412, 1960.
51. V. S. Shpinel', V. A. Bryukhanov, and N. N. Delyagin, Zhur. Eksp. i Teoret. Fiz. 41, 1767, 1961.
52. V. I. Gol'danskii, G. M. Gorodinskii, S. V. Karyagin, L. A Korytko, L. M. Krizhanskii, E. F. Makarov, I. P. Suzdalev, and V. V. Khrapov, Doklady Akad. Nauk SSSR 147, 127, 1962.
53. R. E. Watson, Phys. Rev. 119, 1934, 1960.
54. R. E. Watson and A. J. Freeman, Phys. Rev. 120, 1125, 1960.
55. L. R. Walker, G. K. Wertheim, and V. Jaccarino, Phys. Rev. Lett. 6, 98, 1961.
56. C. H. Townes and B. P. Dailey, J. Chem. Phys. 17, 782, 1949.
57. P. P. Craig, D. E. Nagle, and D. R. F. Cochran, Phys. Rev. Lett. 4, 561, 1960.
58. S. I. Aksenov, V. P. Alfimenkov, V. I. Lushikov, Yu. M. Ostanevich, F. L. Shapiro, and Yen Wu-kuang, Zhur. Eksp. i Teoret. Fiz. 40, 88, 1961.
59. Yu. M. Kagan, Zhur. Eksp. i Teoret. Fiz. 41, 659, 1961.
60. Yu. M. Kagan and V. A. Maslov, Zhur. Eksp. i Teoret, Fiz. 41, 1296, 1961.
61. G. K. Wertheim, Phys. Rev. Lett. 4, 403, 1960.
62. S. L. Ruby, L. M. Epstein, and K. H. Sun, Rev. Sci. Instr. 31, 580, 1960.
63. V. V. Sklyarevskii, B. N. Samoilov, and E. P. Stepanov, Zhur. Eksp. i Teoret. Fiz. 40, 1874, 1961.
64. N. N. Delyagin, V. S. Shpinel', and V. A. Bryukhanov, Zhur. Eksp. i Teoret. Fiz. 41, 1374, 1961.
65. D. Shirley, M. Kaplan, R. W. Grant, and D. A. Keller, Phys. Rev. 127, 2097, 1962.
66. V. A. Bryukhanov, V. I. Gol'danskii, N. N. Delyagin, E. F. Makarov, and V. S. Shpinel', Zhur. Eksp. i Teoret. Fiz. 42, 637, 1962.
67. A. J. F. Boyle, D. S. Bunbury, and C. Edwards, Proc. Phys. Soc. 79, 416, 1962.
68. V. A. Bryukhanov, V. I. Gol'danskii, N. N. Delyagin, L. A. Koryutko, E. F. Makarov, I. P. Suzdalev, and V. S. Shpinel', Zhur. Eksp. i Teoret. Fiz. 43, 448, 1962.
69. G. K. Wertheim, Phys. Rev. 124, 764, 1961.
70. C. Alff and G. K. Wertheim, Phys. Rev. 122, 1415, 1961.
71. D. Shirane, D. Cox, and S. Ruby, Phys. Rev. 125, 1158, 1962.
72. L. M. Epstein, J. Chem. Phys. 36, 2731, 1962.
73. S. V. Karyagin, Doklady Akad. Nauk SSSR 148, 1102, 1963.
74. V. I. Gol'danskii, E. F. Makarov, and V. V. Khrapov, Zhur. Eksp. i Teoret. Fiz. 44, 752, 1963; Phys. Letters 3, 344, 1963.
75. V. I. Gol'danskii, S. V. Karyagin, E. F Makarov, and V. V. Khrapov, Transactions of the Conference on the Mössbauer Effect (Dubna, July 1962), OIYaI (Joint Institute of Nuclear Studies), 1962.
76. Yu. M. Kagan, Doklady Akad. Nauk SSSR 140, 794, 1961.
77. N. E. Alekseevskii, Pham Zuy Xien, V. G. Shapiro, and V. S. Shpinel', Zhur. Eksp. i Teoret. Fiz. 790, 1962.
78. P. Craig, N. Erickson, D. E. Nagel, and R. Taylor, Transactions of the Second Conference on the Mössbauer Effect (Paris, September 1961), Wiley, N.Y.-London, 1962, p. 280.
79. H. Pollak, M. de Coster, and S. Amelinckx, Transactions of the Second Conference on the Mössbauer Effect (Paris, September 1961), Wiley, N.Y.-London, 1962, p. 112.
80. S. S. Hanna, J. Heberle, C. Littlejohn, G. J. Perlow, R. S. Preston, and D. H. Vincent, Phys. Rev. Lett. 4, 177, 1960.
81. G. K. Wertheim and J. H. Wernick, Phys. Rev. 123, 755, 1961.

82. S. de Benedetti, G. Lang, and R. Ingalls, Phys. Rev. Lett. 6, 60, 1961.
83. W. Kerler and W. Neuwirth, Z. Phys. 167, 176, 1962.
84. W. Kerler, Z. Phys. 167, 194, 1962.
85. G. Shirane, W. J. Takei, and S. L. Ruby, Phys. Rev. 126, 49, 1962.
86. J. Solomon, Compt. rend. 250, 3828, 1960.
87. J. Solomon, Compt. rend. 251, 2675, 1960.
88. G. K. Wertheim and J. H. Wernick, Phys. Rev. 125, 1937, 1962.
89. S. Komura, N. Kunitomi, P. Tseng, N. Shikanozo, and H. Takekoshi, J. Phys. Soc. Japan 16, 1479, 1961.
90. E. A. Friedman and W. J. Nicholson, Bull. Am. Phys. Soc. 7, No. 6, 402, 1962.
91. U. Zahn, P. Kienle, and H. Eicher, Z. Phys. 166, 220, 1962.
92. E. Fermi and E. Segre, Z. Phys. 82, 729, 1933.
93. S. A. Goudsmit, Phys. Rev. 43, 636 (1933).
94. É. E. Vainshtein, R. L. Barinskii, and K. I. Narbutt, Zhur. Eksp. i Teoret. Fiz. 23, 593, 1952.
95. R. L. Barinskii, Zhur. Strukt. Khim. 1, 200, 1960.
96. Ya. K. Syrkin, Uspekhi Khim. 31, 397, 1962.
97. E. M. Shustorovich and M. E. Dyatkina, Doklady Akad. Nauk SSSR 128, 1234, 1959.
98. E. M. Shustorovich and M. E. Dyatkina, Doklady Akad. Nauk SSSR 133, 141, 1960.
99. V. A. Bryukhanov, N. N. Delyagin, A. A. Opalenko, and V. S. Shpinel', Zhur. Eksp. i Teoret. Fiz. 43, 432, 1962.
100. A. L. Shawlow, J. Chem. Phys. 22, 1211, 1954.
101. M. M. Yakshin, V. M. Ezuchevskaya, and V. A. Salmenkova, Zhur. Neorg. Khim. 6, 2425, 1961.
102. I. Khaiduk, Uspekhi Khim. 30, 1124, 1961.
103. Ya. K. Syrkin, Uspekhi Khim. 28, 903, 1959.
104. I. P. Gol'dshtein, E. N. Gur'yanova, E. D Deinskaya, and K. A. Kocheshkov, Doklady Akad. Nauk SSSR, 136, 1079, 1961.
105. A. M. Baldin, V. I. Gol'danskii, and I. L. Rozental', Kinematics of Nuclear Reactions, FMgiz, 1959, Sec. 33.
106. D. A. Shirley, M. Kaplan, and P. Exel, Phys. Rev. 123, 816, 1961.
107. D. A. Shirley, Phys. Rev. 124, 354, 1961.
108. A. Yu. Aleksandrov, N. N. Delyagin, K. P. Mintrofanov, L. S. Polak, and V. S. Shpinel', Zhur. Eksp. i Teoret. Fiz. 43, 1242, 1962.
109. A. N. Nesmeyanov, D. I. Mendeleev's Periodic System of the Elements and Organic Chemistry, paper presented at the VIII Mendeleev Congress on General and Applied Chemistry, Izd-vo Akad. Nauk SSSR, 1959.
110. L. A. Blyumenfel'd, V. A. Benderskii, and A. É. Kalmanson, Biofiz. 6, 631, 1961.
111. L. A. Blyumenfel'd, Doklady Akad. Nauk SSSR 148, 361, 1963.
112. L. V. Abramova, N. I. Sheverdina, and K. A. Kocheshkov, Doklady Akad. Nauk SSSR, 123, 681, 1958.
113. J. W. Boyle, S. Weiner, and C. J. Hochanadel, J. Phys. Chem. 63, 892, 1959.
114. O. A. Ptitsyna, O. A. Reutov, and M. F. Turchinskii, Doklady Akad. Nauk SSSR, 114, 110, 1957.
115. A. V. Topchiev, B. A. Krentsel', and L. A. Stotskaya, Uspekhi Khim. 30, 462, 1961.
116. G. A. Abakumov, S. V. Shulyndin, and A. E. Shilov, Kinetika i Kataliz, 1963.
117. W. L. Carrick, K. W. Kluiber, E. F Bonner, L. H. Wartman, F. M. Rugg, and J. J. Smith, J. Am. Chem. Soc. 82, 3883, 1960.
118. M. de Coster and S. Amelinckx, Phys. Lett. 1, 245, 1962.
119. V. I. Gol'danskii and Yu. M. Kagan, Int. T. Appl. Rad. Isotopes, 11, 1, 1961.
120. Yu. M. Kagan and Ya. A. Iosilevskii, Zhur. Eksp. i Teoret. Fiz. 42, 259, 1962.
121. Yu. M. Kagan and Ya. A. Iosilevskii, Zhur. Eksp. i Teoret. Fiz. 44, 284, 1963.
122. A. Yu. Aleksandrov, N. N. Delyagin, K. P. Mitrofanov, L. S. Polak, and V. S. Shpinel', Zhur. Eksp. i Teoret. Fiz. 43, 2074, 1962.
123. A. Yu. Aleksandrov, N. N. Delyagin, K. P. Mitrofanov, L. S. Polak, and V. S. Shpinel', Doklady Akad. Nauk SSSR 148, 126, 1963.
124. R. H. Herber and G. K. Wertheim, Transactions of the Second Conference on the Mössbauer Effect (Paris, September, 1961), Wiley, N.Y.-London, 1962.

125. V. Jaccarino and G. K. Werthein, Transactions of the Second Conference on the Mössbauer Effect (Paris, September 1961), Wiley, N.Y.-London, 1962, p. 260.
126. O. Kistner, V. Jaccarino, and L. Walker, Transactions of the Second Conference on the Mössbauer Effect (Paris, September, 1961), Wiley, N.Y.-London, 1962, p. 264.
127. U. Zahn, P. Kienle, and R. Eicher, Transactions of the Second Conference on the Mössbauer Effect (Paris, September 1961), Wiley, N.Y.-London, 1962, p. 271.
128. N. Kosta, J. Danon, and R. Xavier, J. Phys. Chem. Solids 23, 1783, 1962.
129. J. Holmes and H. Kaesz, J. Am. Chem. Soc. 83, 3903, 1961.
130. G. P. Van der Kelen, Nature 193, 1069, 1962.
131. R. Okawara, Proc. Chem. Soc., p. 383, 1961.
132. D. Alleston, A. Davies, and B. Figgis, Proc. Chem. Soc. p. 457, 1961.
133. I. Beattie, G. McQuillan, and R. Hulme, Chem. and Ind., No. 31, 1429, 1962.
134. G. Van der Kerk, J. Luijten, and M. Janssen, Chimia 16, 10, 1962.
135. J. Luijten, M. Janssen, and G. Van der Kerk, Rec. trav. chim. 81, 202, 1962.
136. M. Janssen, J. Luijten, and G. Van der Kerk, Rec. trav. chim. 82, 90, 1963.
137. H. Clark, R. O'Brien, and J. Trotter, Proc. Chem. Soc., p. 85, 1963.
138. V. A. Bryukhanov, N. N. Delyagin, and Yu. M. Kagan, Zhur. Eksp. i Teoret. Fiz. 45, 1372, 1963.
139. V. A. Bryukhanov, N. N. Delyagin, and Yu. M. Kagan, Zhur. Eksp. i Teoret. Fiz. 46, No. 3, 1964.
140. V. I. Nikolaev and S. S. Yakimov, Zhur. Eksp. i Teoret. Fiz. 46, No. 2, 1964.

NEW RESULTS OF USING THE MOSSBAUER EFFECT IN CHEMISTRY

At the courteous suggestion of Consultants Bureau, which is preparing to publish the English translation of our monograph in the United States, we shall consider briefly certain new results in this supplement. Since it is not the purpose of our monograph to give an exhaustive review of papers on the chemical applications of the Mössbauer effect, neither shall we make any such attempt in this brief supplement.

The chemical elements which have been investigated most often in the past year include iron, tin, iodine, xenon, gold, and several lanthanides. We shall not discuss the interpretation of hyperfine splitting of lines in the Mössbauer spectra of lanthanides, this subject having been considered in detail in R. L. Mössbauer's report to the conference at Cornell University in September, 1963 [1]. In this connection, we note only that A. M. Afanas'ev and Yu. M. Kagan [2] recently framed a theory of the hyperfine structure of lines in paramagnetic substances, in an arbitrary temperature interval.

Among the new results for gold we note [3], in which chemical shifts were observed for dilute solutions of Au^{197} in 19 metals and semiconductors at 4.2°K, and it was shown that as the electronegativity of the solvent atoms decreases, i.e., as $|\psi(0)|^2$ for gold atoms increases (on account of 6s- conductivity electrons), the positive chemical shift increases. This result, which is analogous to that shown in Fig. 20 for tin compounds, attests that ΔR is positive in the case of Au^{197}.

In connection with the considerable attention which the chemical compounds of inert gases are now receiving, an investigation of the Mössbauer spectra of xenon compounds [4], using the isotope Xe^{129} derived from an NaI^{129} source ($T_{1/2} = 1.6 \cdot 10^7$ yr), which was conducted at the Argonne Laboratory in the United States, is of particular interest. The spectra of four compounds were observed: the clathrate complex of Xe with hydroquinone, sodium xenate $Na_4XeO_6 \cdot H_2O$, and the xenon fluorides XeF_2 and XeF_4. In all cases chemical shifts in excess of the experimental error were absent. Since the ratio $\Gamma/E = 2.4 \cdot 10^{-11}$ for Xe^{129}, i.e., is very large, the chemical shift of Mössbauer lines for xenon is only slightly sensitive to changes in $|\psi(0)|^2$. Hence the coincidence of shifts for the clathrate complex, where the xenon atoms are simply held, as such, in an organic cage, and for the chemical compounds of xenon listed above cannot be regarded as proof that the $6s^2$-electrons of xenon persists in these compounds.

The enormous and equal quadrupole splittings found for XeF_4 and XeF_2, where $W = eQq = 1.1 \cdot 10^{-5}$ eV, are much more interesting. The equality of the splittings in the two fluorides may be interpreted as a result of sp^3d^2 hybridization in XeF_4, and sp^3d hybridization in XeF_2 [5]. In the first case (symmetry D_{4h}) the four F atoms lie in the same plane, XOY (sp^2d hybridization), in which eight valence electrons are thus concentrated; two more electron pairs — $p_zd_{z^2}$ — lie on the z axis, and cause the observed splitting. In the second case (symmetry D_qC_{3h}), these two electron pairs ($p_zd_{z^2}$) provide bonds between the Xe and the two F atoms along the z axis, and six electrons (sp^2 hybridization) are concentrated in the XOY plane. In this case, even if the Xe—F bonds are assumed to be wholly ionic, the observed splitting requires a larger quadrupole moment $Q(Xe^{129*})$ than the value Q = 0.12 barn used in calculation. It will be interesting to determine whether both the magnitude and sign of the electronic field gradient in the vicinity of the Xe nuclei are the same in XeF_4 and XeF_2, i.e., to compare the role of p_z electrons and p_z holes in the two cases. Here we note an earlier paper [4] on observation of the Mössbauer effect for Kr^{83} in the clathrate complex of krypton and hydroquinone [6]. Further work with Kr^{83} is very urgent in connection with the latest advances in synthesizing chemical compounds of krypton (see, for example, [7]).

An investigation of the Mössbauer effect in certain iodine compounds [8], using the radioactive isotope I^{129} ($1.6 \cdot 10^7$ yr) as absorber, and β-active Te^{129} (70 min) as radiator, gave curious results. The chemical shift for the iodine compounds studied was as follows:*

Compound	KIO_3	NH_4IO_3	KI	KIO_4
δ (cm/sec)	$+0.16 \pm 0.02$	$+0.13 \pm 0.03$	-0.052 ± 0.007	-0.234 ± 0.01

At first glance this sequence may seem strange. However, it has a simple and quite natural explanation [8]. In KI, where the bond is predominantly ionic, the outer shell of the iodine atom has the electronic structure $5s^2p^6$. In iodates with the perovskite-type structure, the I atoms lie at the centers of the octahedra and are surrounded by six O atoms, so that the six I—O bonds are directed at 90° angles to one another. It may be concluded from this that the covalent bonds here are due mainly to 5p-electrons whose removal from the iodine atoms decreases the screening of the 5s-electrons and increases $|\psi(0)|^2$. On the other hand, $|\psi(0)|^2$ is decreased in the periodate KIO_4, since strong sp^3 hybridization occurs here; when the tetrahedral IO_4^- group having four I—O bonds is formed, six 5p- and two 5s-electrons are removed from each iodine atom.

Thus, when $|\psi(0)|^2$ is increased, the chemical shift becomes more positive, as in Sn^{119} and Au^{197}; this indicates that ΔR has a positive sign. The spectra of both the iodates investigated in[8], contrary to those of KI and KIO_4, exhibit quite strong quadrupole splitting which, in the case of I^{129}, is caused by the presence of a quadrupole moment in both the excited $(+\frac{5}{2})$ and ground $(+\frac{7}{2})$ states; this leads to splitting into eight lines. For the ground state of I^{129} in KIO_3, $W = eQq \cong 2.7 \cdot 10^{-6}$ eV; for the excited state, $W^* = eQ^*q \cong 3.35 \cdot 10^{-6}$ eV. This result was used in [8] to determine $Q = -0.69$ barn, but not for interpretation from the viewpoint of the structure of KIO_3. In this connection it would be desirable to compare the quadrupole splittings for iodates with different cations, e.g., to compare KIO_3 with $LiIO_3$, since these cations have different coordination numbers (six O atoms for Li and twelve for K) [9].

The Mössbauer spectra of tin have been further investigated in our laboratory. A quite detailed analysis of organotin oxides $(R_2SnO)_n$, for which a polymeric structure of the type

$$
\begin{array}{ccccc}
& R & & R & R \\
& | & & | & | \\
-O-&Sn&-O-&Sn&-O-Sn- \\
& | & & | & | \\
& R & & R & R \\
\end{array}
$$

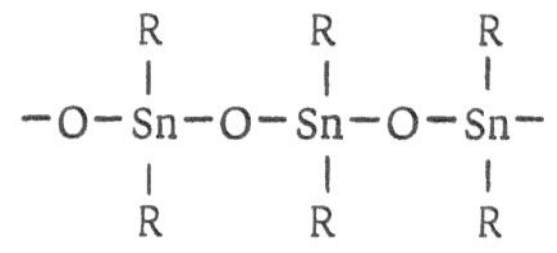

was proposed earlier, was carried out [10]. Comparison of the spectra of these oxides ($\delta \approx -1.7$ mm/sec relative to β-Sn; $\Delta = \frac{1}{2}W = 1.7$-2.1 mm/sec) with those of the carboxylates $R' - C - O - Sn - O - C - R'$ (with R, O, R substituents) and bis-alkyl-substituted tin oxides of the type

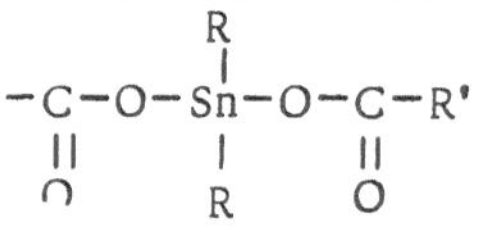

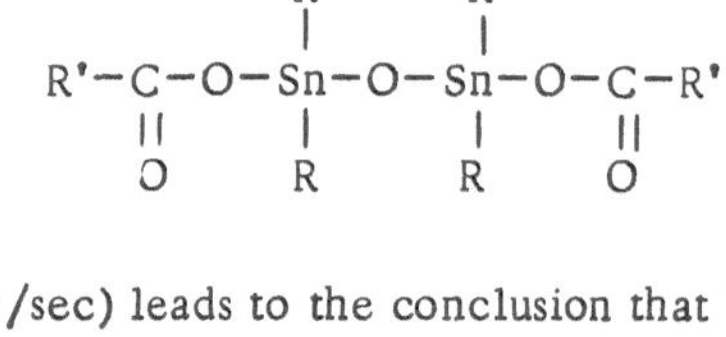

$$
\begin{array}{ccccc}
& R & & R & \\
& | & & | & \\
R'-C-O-&Sn&-O-&Sn&-O-C-R' \\
\| & | & & | & \| \\
O & R & & R & O \\
\end{array}
$$

($\delta \approx -1.35$ mm/sec; $\Delta = 3.2$-3.5 mm/sec) leads to the conclusion that both the gradient q and $|\psi(0)|^2$ for Sn nuclei in the oxides R_2SnO are decreased (q very substantially). It might be supposed that in all the above

---

*In [8] the sign convention for δ is opposite to that adopted in our monograph: the + sign corresponding to motion of the source and absorber away from each other; here the signs of δ have the same significance as in our main text (see p. 15).

compounds the nearest bonds of the tin atoms are identical: $-O-\overset{\overset{C}{|}}{\underset{\underset{C}{|}}{Sn}}-O-$; however, in those cases where oxygen atoms are followed by carbonyl groups, a very strong inductive effect is observed which sharply increases the degree of ionic character of the Sn–O bonds, and hence also the quadrupole splitting. In this case, however, one might expect $|\psi(0)|^2$ in R_2SnO not to decrease but, on the contrary, to increase in comparison with the other compounds named.

The explanation of the above results by the inductive effect is contradicted by comparison of the spectra for a series of compounds having four Sn–C bonds, namely, R_3SnR', where R = Alk and R' = Vin, Ph, CH_2COCH_3, and CH_2Cl.

In the indicated series, the Taft constants characterizing the magnitude of the inductive effect are identical (≈0.6) for the first three substituents R', whereas the inductive effect for the $-CH_2Cl$ group is stronger than for the $-CH_2COCH_3$. However, $|\psi(0)|^2$ values, approximately the same for R = Vin and Ph (δ = –1.6 mm/sec), proved to be larger for R' = CH_2COCH_3 (δ = –1.38 mm/sec) and especially for R' = CH_2Cl (δ = –1.27 mm/sec), a sequence which again is opposite to that which would be expected from the inductive effect. In contrast to all compounds having four Sn–C bonds, quadrupole splitting (Δ = 1.13 mm/sec) was observed in the case of triethyltin acetonate. We explained all the above facts on the hypothesis that tin atoms have coordination number 5 in organotin oxides and trialkyltin acetonates (see footnote on p. 36). The appearance of sp^3d hybridization leads to decrease in $|\psi(0)|^2$ because of the screening action of the 5d orbital; the quadrupole splitting proves to be less in this case than in the case of four sp^3 bonds differing strongly in ionic character but, of course, is more than in the case of four identical sp^3 bonds. The proposed structure of the tin oxides R_2SnO and the trialkyltin acetonates $R_3SnCH_2COCH_3$ is shown in Fig. S-1.

It should be noted, however, that the given interpretation is not the only one possible. In the case of R_2SnO, as well as for many other tin compounds, the possibility of the formation of additional d_π-p_π bonds with participation of free d orbitals of the tin must be taken into account. We are now conducting a series of additional experiments in order to find out the possible role of such bonds in quadrupole splitting and chemical shifts of the Mössbauer spectra of tin compounds.

The spectra of the following organic compounds of bivalent tin: Ph_2Sn and Bu_2Sn ($\delta \approx$ –1.2 mm/sec; Δ = 0), which we studied together with V. Ya. Rochev and V. V. Khrapov, were very similar to those observed for SnR_4 and $R_3Sn-SnR_3$, where R = Ph, Bu, Et. Apparently a polymer $(R_2Sn)_n$ having Sn–Sn and Sn–C bonds is formed here. Analogously to $R_3Sn-SnR_3$, SnR_2 compounds are easily oxidized in air; the corresponding change in the Mössbauer spectrum of $(R_2Sn)_n$ with formation of a doublet enabling one to identify the oxidation product as $(R_2SnO)_n$ is shown in Fig. S-2. Based on the data of Fig. S-2, the entire kinetic oxidation curve may be plotted.

In connection with the similarity of the Mössbauer spectra of SnR_2 and SnR_4, the contradictory character of the data on spectra of inorganic compounds of bivalent tin must be noted again, particularly the position of the $SnBr_2$ singlet line (see p. 33), found recently by another author [11], which is quite at variance with the aggregate of such data. The question arises as to whether polymers are formed in these cases, the tin here too actually being quadrivalent.

Later we investigated a number of tin halides and their complex compounds [12]: K_2SnF_6, Cs_2SnF_6, and more than 30 complexes of $SnCl_4$, $SnBr_4$, and SnI_4 with ethylenediamine (En), tetra- and hexamethylenediamines, pyridine, piperidine, tetrahydrofuran, and tetrahydrothiophene.

Visible quadrupole splitting was absent in all cases, as for $SnHal_4$ (except SnF_4, for which Δ = 1.66 mm per sec), although line broadening was observed in a number of cases. The chemical shifts indicated a decrease in $|\psi(0)|^2$ on complex formation, lessening in the order I, Cl, Br, F. Thus, for instance, in the complexes $SnX_4 \cdot 2En$ the value of δ was about the same ($\delta \approx$ –2.2 mm/sec) for X = Cl, Br, and I, the changes in δ in comparison with SnX_4 being equal to 0.2, 1, and 1.2 mm/sec, respectively. These data are interpreted in [12] as a result of screening of the 5s-electrons on filling of the 5d orbitals in the tin atoms.

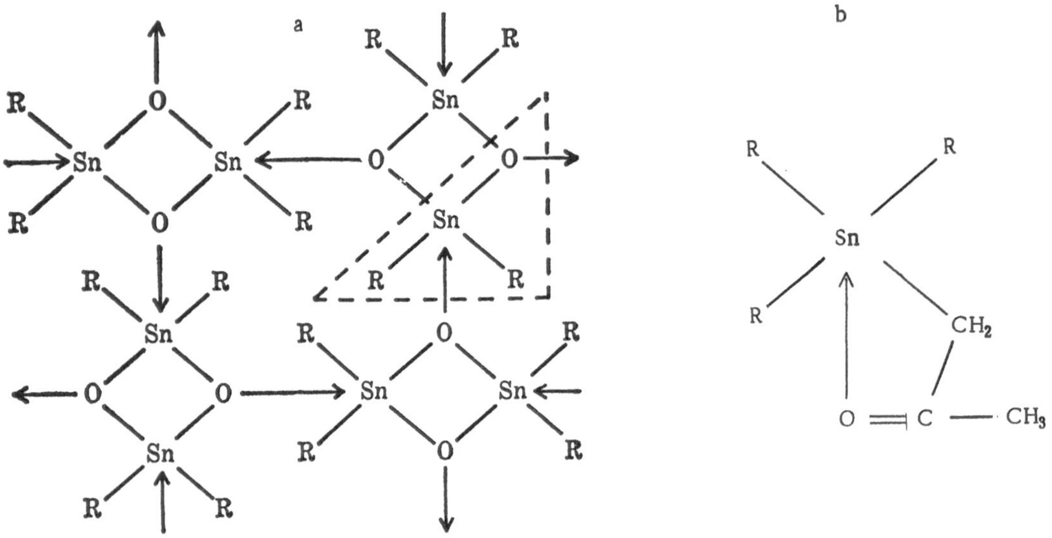

Fig. S-1. Supposed structures of polymeric organotin oxides $(R_2SnO)_n$ (a) and trialkyltin acetonates $R_3SnCH_2\underset{\overset{\|}{O}}{C}-CH_3$ (b).

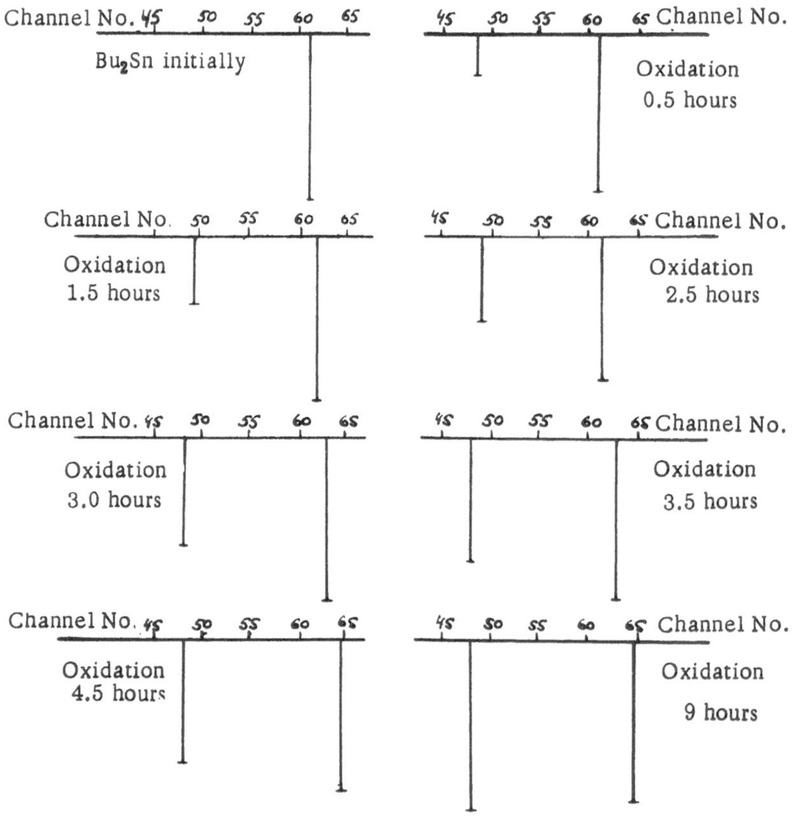

Fig. S-2. Position of lines in Mössbauer spectra of a given sample of Bu_2Sn with respect to the duration of its oxidation in air. (Magnitude of the effect is proportional to length of line.) Channel width 0.124 mm/sec. Zero velocity relative to SnO_2 corresponds to Channel No. 49.5.

However, the most characteristic distinguishing feature of all compounds of tin with coordination number 6 — SnO_2, SnF_4, and the above-mentioned complexes — proved to be the especially high probability of the Mössbauer effect at 78°K and the persistence of a fully observable effect at room temperature. It was found in [10] that while the quantities δ and Δ for the compounds (para-X-$C_6H_4)_2SnO$ do not vary, the Debye-Waller factor f' at 78°K is more than doubled in the series X = H, Cl, Br, I, i.e., as the molecules become heavier.

The data of [12] showed that values of f' and their temperature dependence, and not the chemical shift or quadrupole splitting, are sometimes the most reliable basis for identifying the structure of compounds and the character of chemical bonds; for instance, the appearance or disappearance of the Mössbauer effect at room temperature may enable one to observe the kinetics of various conversions in which compounds of tin with coordination number 6 are formed or decomposed. Based on our data, we must include, for instance, the tetrapropiolate $[HC \equiv CCOO]_4Sn$ and trinitromethyltin chloride $ClSn(CH_2NO_2)_3$ among such compounds. It would be very desirable to conduct an x-ray structural investigation of the substances named.

The qualitative interpretation of the results is based on the decrease in the mean square amplitudes of thermal vibration ($\overline{x^2}$) of Sn atoms in the lattice with increase in total strength of all their bonds (sp^3d^2 instead of sp^3), and upon the increase in the number of high-frequency optical branches in the vibration spectra of lattices having polyatomic unit cells.

Values of chemical shift and quadrupole splitting obtained from Mössbauer spectra cannot give reliable data on the ionic character and hybridization of chemical bonds until an exact theory is framed connecting the density of s-electrons in the nucleus ($|\psi_s(0)|^2$) and the electric field gradient (q) with the characteristics of the chemical bond. Hence it is desirable to seek rules to establish relations of some kind between Δ and δ in various classes of compounds containing Mössbauer nuclei.

The presence of a definite correlation between δ and Δ for the particular case of A_iSnB_{4-i}-type compounds with an arbitrary degree of ionic character (i) and hybridization (α) of the Sn—A and Sn—B bonds was demonstrated by semiempirical calculations performed by our co-workers E. F. Makarov and O. V. Rostovskii.

The indicated correlation is a simple consequence of the character of the wave functions of the four valence electrons, and for $ASnB_3$ has the form:

$$\delta = \delta_0 \left\{ K \left[1 - i_{Sn-A} + \frac{\Delta}{\Delta_0} \right] - 1 \right\},$$

where

$$\delta_0 = 1.55 \cdot 10^{-23} \frac{\Delta R}{R} \ |\psi_s(0)|^2_{source} \frac{mm}{sec}; \qquad K = \frac{|\psi(0)|^2_{abs}}{|\psi(0)|^2_{source}},$$

$\Delta_0 = \frac{1}{2}eq_0Q$ is the quadrupole splitting on removal of one p_z electron, and $q_0 = 9 \cdot 10^{18}$ V/cm^2.

As a standard from which to compute chemical shifts α-Sn (i = 0; Δ = 0) is chosen. On the hypothesis that the difference in the position of the α- and β-Sn lines (0.6 mm/sec) is determined mainly by their different densities, one can find the constant δ_0, equal to 4.35 mm/sec in the case of α-Sn.

The following comparison of δ for $SnHal_4$ with the i_{Sn-Hal} values derived in [13] from experiments in nuclear quadrupole resonance (with corrections for d_π-p_π bonds) gives the value of K for all these compounds and also, with certain additional assumptions, for Ph_3SnHal.

The quadrupole splitting of lines for Ph_3SnHal is equal to $\Delta = \Delta_0 \cdot 3\alpha\,(i_{Sn-Hal} - i_{Sn-Ph})$, and hence, after calculating the difference in the degree of ionic character of Sn—Hal and Sn—Ph bonds according to [13], one can determine the quadrupole moment $Q(Sn^{119*})$ with respect to α, from α = 0.25 (sp^3 hybridization) to α = 0.33 (sp^2 hybridization for Sn—Ph bonds). Such an estimate gave the extreme limits Q = (9.8-12.8)$\cdot 10^{-26}$ cm^2.

The ratio $\Delta R/R(Sn^{119})$ also was recalculated from the beginning. Calculation according to the Fermi-Segrè method gave $|\psi(0)|^2_{5s} = 1.6 \cdot 10^{26}$ cm^3; instead of white (β) tin, gray (α) tin was now chosen as a standard with four purely covalent sp^3 bonds. This resulted in $\Delta R/R = 1.4 \cdot 10^{-4}$

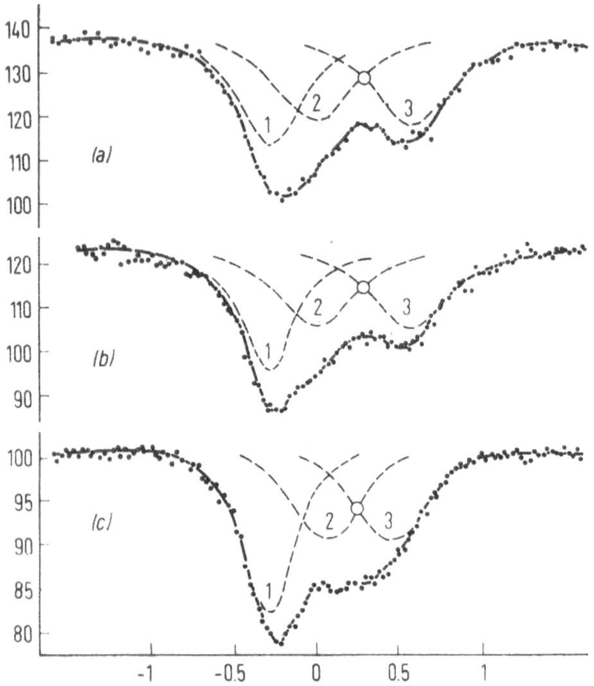

Fig. S-3. Mössbauer spectra of certain iron ferro- and ferricyanides: a) Prussian blue; b) Turnbull's blue; c) soluble Prussian blue $\{KFe[Fe(CN)_6]\}$.

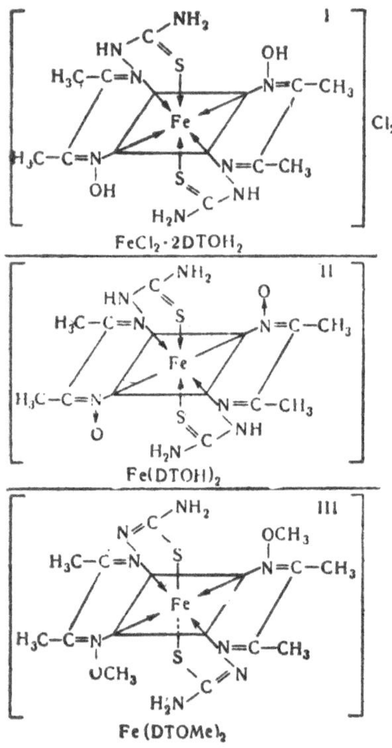

Fig. S-4. Structural formulas of complex compounds of iron (P) with diacetyloxime thiosemicarbazone and its derivatives.

The largest number of investigations of the chemical profile during the past year were conducted, as before, with the isotope Fe^{57}. A very detailed analysis of the corresponding results is given in papers by the German group of investigators [14, 15]. There would be little point in restating the entire contents of these papers here, and hence we shall merely discuss the main data relatively briefly. We shall first treat the investigations of the structure of complex cyanides containing iron, both in the form of a complex-former and in the outer coordination sphere.

As mentioned in Chapter V, the Mössbauer spectra of low-spin ferri- and ferrocyanide complexes differ substantially from the spectra of simple Fe^{III} and Fe^{II} salts. The presence of quadrupole splitting of lines for the paramagnetic anion $[Fe^{III}(CN)_6]^{3-}$ enables one to distinguish its spectrum from the singlet line of the diamagnetic anion $[Fe^{II}(CN)_6]^{4-}$. On the other hand, it is still easier to differentiate the spectra of the Fe^{++} and Fe^{+++} cations (see Figs. 15 and 16).

Owing to this, it was clearly demonstrated in [16, 17] that the reaction of Fe^{III} with potassium ferrocyanide $K_4Fe(CN)_6$, and that of Fe^{II} with potassium ferricyanide $K_3Fe(CN)_6$ give identical products, whose spectra can, as is evident from Fig. S-3, be distinguished by the singlet line of the $[Fe^{II}(CN)_6]^{IV}$ anion and the doublet of the Fe^{III} cation.

It was proved that Prussian blue (Fig. S-3a) and Turnbull's blue (Fig. S-3b) are the same compound $Fe_4^{III}[Fe^{II}(CN)_6]_3$ [16] and that soluble Prussian blue $KFe[Fe(CN)_6]$ (Fig. S-3c) is the compound $KFe^{III}[Fe^{II}(CN)_6]$ no matter how it is prepared [15], [16], i.e., that the reaction $Fe^{II} + [Fe^{III}(CN)_6]^{IV} \rightarrow Fe^{III} + [Fe^{II}(CN)_6]^{IV}$ is completely shifted to the right. Thus the intense color of Prussian blue cannot be explained as the oscillation of charge between the outer and inner iron atoms.

Results analogous to those of [16, 17], were obtained in our laboratory and were under discussion when the authors of [17] kindly sent out a report of their results. For Cu^{II} and Ag^{I}, contrary to iron, there exist complexes both with the $-[Fe^{II}(CN)_6]^{IV}$ and $[Fe^{III}(CN)_6]^{III}$ anions [17]; we observed the same picture in the case of Sn^{IV}.

In the formation of the complex $[Fe^{II}(CN)_6]^{IV}$ the six σ-bonds between the iron and the cyanide groups (owing to $3d^24sp^3$ hybridization) might bring the formal charge on the Fe down to −4. Actually, however, the effective charge on the iron in

TABLE S-I

Compound*	Ligand	δ, mm/sec (−130°C)	Δ, mm/sec (−130°C)	Sign of q	π-bond strength
$Na_2[Fe(CN)_5NO]$	NO^+	−0.13	1.73	+	↑
$K_4[Fe(CN)_6]$	CN^-	+0.10	0	0	
$Na_5[Fe(CN)_5SO_3]$	SO_3^{2-}	+0.11	0.73		
$Na_4[Fe(CN)_5NO_2]$	NO_2^-	+0.15	0.86	−	
$Na_3[Fe(CN)_5NH_3]$	NH_3	+0.15	0.68	−	

*Number of molecules of water of crystallization not given.

ferrocyanides is evidently close to the value η = +1 observed for ferricyanides (see p. 28). One reason for the decrease in the number of electrons near the iron atoms is the appearance of additional π-bonds, i.e., reverse donor-acceptor bonds involving the three d orbitals of iron which do not take part in σ-bonds. By itself, this effect could change η from −4 to −1; the further change in η by two units is evidently due to the partial ionic character of the bonds between Fe and CN^-.

The data of [14, 17] on the spectra of various prussiates — complexes of the type $[Fe^{II}(CN)_5X]$, where X = NO^+, SO_3^{2-}, NO_2^-, NH_3, and also CN^- — are of substantial interest for understanding the role of π-bonds. The stronger the π-bond with the ligand X is, and hence the more intensely the 3d electron cloud is withdrawn from the iron atom, the less the screening effect of the inner s electrons must be, and the larger $|\psi(0)|^2$ is. The corresponding relation between π-bond strength and chemical shift relative to iron in stainless steel is illustrated by Table S-I.

Unilateral strengthening or weakening of π-bonds in comparison with the spherically symmetrical anion $[Fe^{II}(CN)_6]$ gives rise to quadrupole splitting whose sign, according to the authors of [14, 15], is different in different prussiates. The possibility of inversion of the quadrupole splitting as a result of very slight changes in molecular structure, noted in [14, 15] and our papers [18, 19], is a very important fact. Direct study of such inversion, e.g., by measuring the spectra in external magnetic fields, may provide much new information for ascertaining the character of all kinds of changes in the intramolecular distribution of electron-cloud densities.

It is especially desirable to determine directly the sign of quadrupole splitting, since no indirect line of reasoning is fully reliable. Thus, for instance, measurement of the effective charge of Fe atoms in cyanide complexes and carbonyls shows that this charge not only decreases in absolute value owing to π-bonds, but also changes sign, contrary to expectation — it becomes positive. The possibility of asymmetry of the two quadrupole-splitting peaks, even in perfectly isotropic polycrystalline samples, as a result of anisotropy of thermal vibrations in the corresponding single crystals, complicates the choice of sign of the gradient q, which depends on which of the peaks is smaller. On the other hand, the change in character of the indicated asymmetry in polycrystalline samples may be directly connected with inversion of quadrupole splitting, as we assume, e.g., for the case of Ph_3SnCl and $PhSnCl_3$ (see Fig. 21).

Many papers [14, 15, 20, 21] are devoted to the Mössbauer spectra of iron carbonyl $Fe(CO)_5$ and various derivatives of it. As an example we give the results of [21] in Table S-II.

The carbonyl $Fe(CO)_5$ is a trigonal bipyramid; five Fe−Co bonds are provided here by $3d_{z^2}4sp^3$ hybridization; the other four 3d orbitals of the iron take part in d_π-p_π bond formation. These bonds may decrease the formal negative charge on the iron atom from −5 to −1; for the further change to η_{Fe} = +0.4 (see p. 28) the bonds must be partially ionic.

The nonacarbonyl $Fe_2(CO)_9$ contains carbonyl groups of two kinds: three such groups form bridging bonds between two iron atoms; the rest give three more Fe−CO bonds for each Fe atom. For the dodecarbonyl $Fe_3(CO)_{12}$ three types of structure were discussed — one based on a triangle of three equivalent iron atoms, a second, with bridging bonds, of the type (3; 3; 3; 3):

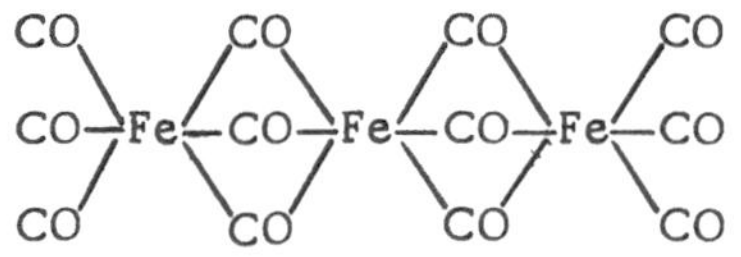

and a third, with bridging bonds, of the type (4; 2; 2; 4).

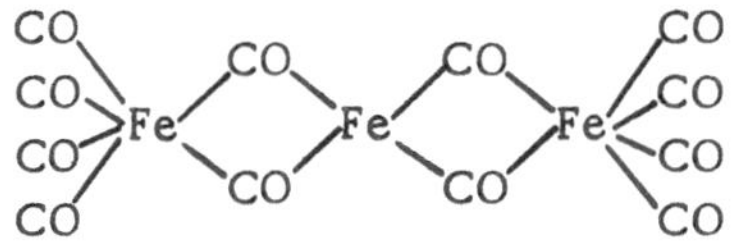

The presence of two components in the Mössbauer spectrum (doublet a and singlet b, see Table S-II) eliminates the first structure; however, it does not provide an unambiguous choice between the other two: the authors of [14, 15] are inclined toward (4; 2; 2; 4); the authors of [21] favor (3; 3; 3; 3). Data on chemical shifts and, to a lesser degree, quadrupole splittings of carbonyl derivatives of iron are interpreted in these papers on the basis of concepts regarding the effect of the 3d- and 4s-electrons of iron on $|\psi(0)|^2$, and the formation of additional π-bonds. In this case different interpretations of identical results are proposed in many instances; thus the iron bonds in $Fe(CO)_4I_2$ are treated as d^2sp^3 in [21] and as dsp^3 in [15], it being assumed that the sign of the field gradient in the vicinity of the iron nucleus, contrary to the case of iron carbonyl, is negative here.

It is obviously desirable to accumulate further experimental data on Mössbauer spectra, particularly with regard to the sign of quadrupole splitting, as well as additional data, e.g., data relative to the effective charges of iron atoms in various compounds. $Fe_2(CO)_9$ and $Fe_3(CO)_{12}$, which are examples of compounds whose spectra are different despite the presumable identity of the nearest neighbors of the iron atoms (six carbonyl groups), demonstrate the appreciable effect of changes in "remote" bonds on the electron shells of the iron itself. We made an analogous observation in [18] for complex compounds of Fe^{II} with diacetyloxime thiosemicarbazone, shown in Fig. S-4. The spectral characteristics of these compounds are given in Table S-III.

In passing from compound I to P, two protons are removed from the NOH groups; in passing from I to Sh, two protons are removed from the NH groups. Wave functions approximating a linear combination of atomic orbitals (LCAO), based on a scheme of possible molecular σ-orbitals for the symmetry D_{2h}, were given in [19], and an approximate expression was found for the electric field gradient at the Fe^{57} nucleus:

$$q \approx [(\alpha_2^2 - \alpha_3^2) + (\alpha_2^2 - \alpha_4^3)]\, q_{4p_z},$$

where α_2, α_3, and α_4 are LCAO coefficients for the atomic ($4p_x$, $4p_y$, and $4p_z$) functions of iron, and q_{4p_z} is the gradient produced by one excess $4p_z$ electron.

Knowing the directions of inductive effects (i.e., the directions of changes in the quantities α_i^2) and having determined experimentally the change in q on passing from I to P and Sh, we could determine the sign of q

TABLE S-II

Compound	T, °K	δ relative to Fe in stainless steel, mm/sec	Δ, mm/sec	Half-width of peak, mm/sec
$Fe(CO)_5$	78	+0.035 ± 0.043	2.57 ± 0.04	0.43 ± 0.04
$Fe_2(CO)_9$	78	+0.232 ± 0.043	0.541 ± 0.043	0.43 ± 0.04
$Fe(CO)_{12}$	78 a	+0.230 ± 0.043	1.115 ± 0.043	0.365 ± 0.043
	b	+0.158 ± 0.043	0	0.48 ± 0.04
$Fe(CO)_4I_2$	78	+0.50 ± 0.08	0.85 ± 0.08	0.636 ± 0.08
	298	+0.49 ± 0.08	0.72 ± 0.08	0.76 ± 0.12

TABLE S-III*

Compound	δ, mm/sec		Δ, mm/sec	
	78°K	300°K	78°K	300°K
$FeCl_2 \cdot 2DTOH_2$ [I]	0.41	0.36	0.65	0.66
$Fe[DTOH]_2$ [P]..........	0.23	0.16	2.02	2.02
$Fe[DTOMe]_2[Sh]$........	0.44	0.39	0.88	0.88

*The error in δ and Δ is ±0.01 mm/sec; δ values are given relative to iron in stainless steel.

and estimate the inductive effect from the differences in the α_i^2. On the assumption that on passing from I to P, removal of protons from the NOH groups shifts the electronic charge toward the iron through the nitrogens, we obtained $q_{II} < 0$. On passing from I to Sh this charge is shifted toward the Fe through the sulfurs, which leads to the conclusion: $q_{III} > 0$. Further considerations, connected with the character of asymmetry of the quadrupole-splitting peaks and the magnitudes of the inductive effects, support the conclusion that $q_I < 0$.

Thus the hypothesis was advanced in [19] that not only direct substitution of ligands, as in prussiates, but also remote rearrangements may cause inversion of quadrupole splitting. It is a primary problem to find out directly whether such inversion occurs.

Experiments with the isotope Fe^{57} gave a number of other new results of interest for the problems dealt with in our monograph.

The Mössbauer spectra for Fe^{III} acetylacetonate (source: Co^{57} in chromium) and $K_4Fe(CN)_6$ (source: Co^{57} in Co^{III} acetylacetonate) were compared in [22]. In the first case a single line was obtained; in the second, a complex spectrum caused by various chemical consequences of K-electron capture by the Co^{57} nucleus. Although the formation of iron in the form of Fe^{II} predominated in the case of CoO (see p. 22), the appearance of the state Fe^{III} proved to be much more probable in the present case than the appearance of Fe^{II}. As in the case of CoO, the formation of higher-valence states of iron was not observed in [22].

Curious results were obtained by the authors of [23], in which the Mössbauer effect was investigated for Fe^{57} atoms carried on the surface of Al_2O_3. Strong asymmetry of quadrupole splitting was observed at 300°K but was absent at 77°K. The authors accepted our explanation of this asymmetry as resulting from the anisotropy of the thermal-vibration amplitudes of surface atoms (see p. 37). In the given case this explanation made it possible to estimate the absolute value of the difference in the squares of thermal-vibration amplitudes in the directions parallel and perpendicular to the surface. After determining the Debye-Waller factor it will be possible to obtain each of the mean-square amplitudes separately and to calculate the force constants for the surface atoms.

The authors of [24] explain the pronounced asymmetry of quadrupole splitting, noted by them in the case of tellurium, also as caused by the anisotropy of thermal-vibration amplitudes.

In concluding our description of new experiments with Fe^{57}, observations of the Mössbauer effect in a viscous liquid [25, 26], namely, glycerol (see p. 43), must be noted.

The temperature dependence of the effect was studied in [25] in the case of an absorber consisting of iron sulfate solutions in glycerol in the interval from −100 to +10°C. As the temperature rose from −100 to 0°C, the line width increased by an order. The results were used to calculate the dependence of the mean-square shift of the Fe^{3+} and Fe^{2+} ions on diffusion time. It was found that the shift takes place especially rapidly, virtually instantaneously, during the first 10^{-8} sec, after which it proceeds much more slowly. In this case the initial shift of Fe^{2+} ions is especially rapid; the rates of shift for Fe^{2+} and Fe^{3+} subsequently become approximately equal.

The authors of [26] dissolved a source – Co^{57}, used in the form of $CoCl_2$ and also introduced into porphyrin and hemoglobin molecules – in glycerol. Metallic iron and $K_4Fe(CN)_6$ were used as absorbers. A comparison of the temperature dependence of line width with theory [27] was used to determine the radius (a) of the diffusion molecules. The value a = 0.65 ± 0.14 A was found for $CoCl_2$; for hemoglobin a = 7.4 ± 2.5 A, which is far less than the value obtained from the diffusion coefficient in water at 20°C (27 A) and may attest to partial dissociation of hemoglobin molecules under the conditions of the experiments being described.

As we mentioned above, together with the chemical shift and quadrupole splitting of Mössbauer spectrum lines, the value of the Debye-Waller factor and its temperature dependence often prove to be important specific characteristics. Another essential characteristic is the Mössbauer-line width. Yu. M. Kagan's papers [28, 29] on the theory of this width and the general theory of hyperfine structure, are noted in this connection.

The question of Mössbauer-line broadening and the relation between this broadening and the excitation spectra in crystals, temperature, and character of nuclear transition arose at the very beginning of investigations of the Mössbauer effect. The possibility of using study of line width and shape as an independent source of information regarding a substance, as well as the problem of obtaining very narrow lines, is obviously of particular interest here. The first adequate solution of this problem is given in the papers cited above.

The question of Mössbauer-line broadening due to thermal red shift is considered in [28, 29]. The author found the exact line shape and connected this line broadening with the vibration spectrum for both ideal and nonideal crystals. In this case it was shown, contrary to the early papers, that residual broadening exists even in strictly regular crystals. Although this broadening is slight for ordinary lines as a rule, it exceeds the natural width by four orders at $T \sim \theta$ in the case of long-lived isotopes such as those of silver. The broadening under consideration is increased in crystals having narrow optical vibration bands, and becomes especially large when the Mössbauer radiator is a light impurity atom in a heavy matrix and vibrates to a substantial degree of discrete frequencies. This broadening decreases sharply with falling temperature and approaches zero as $T \rightarrow 0$.

Yu. M. Kagan and A. I. Afanas'ev [2] developed a general graphical method which enables one to find the γ-quantum absorption (emission) spectrum of a crystal in the presence of hyperfine interaction. The problem of the width of individual lines of hyperfine structure in a ferromagnetic material was solved by this method [28]. Contrary to the general accepted viewpoint, it was found that there is always some residual broadening, which is due primarily to fluctuations of the z component of electron spin. If the gap in the spin-wave spectrum is small enough, so that there are excitations with frequencies of the order of the hyperfine splitting, still another broadening mechanism exists, connected with real transitions between sublevels of the hyperfine structure. Both types of broadening become more pronounced as the value of hyperfine splitting increases and the characteristic frequencies of the spin-wave spectrum decreases.

The case where the hyperfine structure has a purely quadrupole character presents an analogous picture. Now the broadening results from fluctuations of hyperfine interaction due to vibration of the atoms in the crystal.

The general theory of the hyperfine structure of Mössbauer lines in paramagnetic materials also was worked out in [21]. On the hypothesis that electron-spin relaxation has a spin-lattice character, the authors obtained results which are valid for an arbitrary ratio of relaxation frequencies to characteristic frequencies of hyperfine interaction. The interesting and rather unexpected character of the temperature dependence of hyperfine structure in paramagnetic materials, which was found in this paper for a number of cases, should be especially noted.

In conclusion I would like to address a few words to my readers. On looking over my monograph after publication (always done with fresh perceptions) and preparing it for translation, I found many annoying errors. No doubt there are even more errors which I failed to note. Hence, I repeat these words of the preface: I shall be exceedingly grateful to all who read this booklet attentively and inform me of any shortcomings they note.

LITERATURE CITED

1. R. L. Mössbauer, General Aspects of Nuclear Hyperfine Interaction in Salts of the Rare Earths. III. Conference on Mössbauer Effect, Cornell University, September, 1963.
2. A. M. Afanas'ev and Yu. M. Kagan, Zhur. Eksp. i Teoret. Fiz., 45, 1660 (1963).

3. P. H. Barrett, R. W. Grant, M. Kaplan, D. A. Keller, and D. A. Shirley, UCRL – 10608. February (1963).
4. C. L. Chernick, C. E. Johnson, Y. G.Malon, G. Y. Perlow, and M. R. Perlow, Phys. Rev. Lett., 5, 1031, (1963).
5. S. Yamada, Rev. Phys. Chem. Japan 33, 39, (1963).
6. Y. Mozoni, P. Hillman, M. Pasternak, and S. Ruby, Phys. Rev. Lett., 2, 337, (1962).
7. A. V. Grosse, A. D. Kirshenbaum, A. G. Streng, and L. V. Streng, Science 139, 1047, (1963).
8. H. De Waard, G. De Pascuali, and D. Hafemeister, Phys. Rev. Lett., 5, 217, (1963).
9. B. F. Ormont, Structures of Inorganic Substances [in Russian], Moscow, Gostekhizdat, (1950).
10. V. I. Gol'danskii, E. F. Makarov, R. A. Stukan, V. A. Trukhtanov, and V. V. Khrapov, Doklady Akad. Nauk SSSR, 151, 357, (1963).
11. V. A Bukarev, Zhur. Eksp. i. Teoret. Fiz., 44, 852, (1963).
12. V. I. Gol'danskii, E. F. Makarov, R. A. Stukan, T. N. Sumarokova, V. A. Trukhtanov, and V. V. Khrapov, Doklady Akad. Nauk SSSR, 157, (1964).
13. M. A. Whitehead and H. H. Jaffe, Theoret. Chim. Acta (Berl.), 1, 209, (1963).
14. E. Fluck, W. Kerler, and W. Neuwirth, Angew. Chem., 75, 461, (1963).
15. W. Kerler, W. Neuwirth, and E. Fluck, Z. Physik., 175, 200, (1963).
16. J. F. Duncan and P. W. R. Wigley, J. Chem. Soc., 1120, (1963).
17. W. Kerler, W. Neuwirth, E. Fluck, P. Kuhn, and B. Zimmermann, Z. Physik., 173, 321 (1963).
18. A. V. Ablov, G. N. Belozerskii, V. I. Gol'danskii, E. F. Makarov, V. A. Trukhtanov, and V. V. Khrapov, Doklady Akad. Nauk SSSR, 151, 1352, (1963).
19. A. V Ablov, I. B. Bersuker, and V. I. Gol'danskii, Doklady Akad. Nauk SSSR, 152, 1391, (1963).
20. M. Kalvins, U. Zahn, P. Kienle, and K. Eicher, Z. Naturforsch., 17^a, 494, (1962).
21. R. H. Herber, W. R. Kingston, and G. K. Wertheim, Inorg. Chem., 2, 153, (1963).
22. G. K. Wertheim, W. R. Kingston, and R. H. Herber, Z. Chem. Phys., 37, 687, (1962).
23. P. A. Flinn, S. L. Ruby, and W. L. Kehl, "The Mössbauer Effects for Surface Atoms."
24. E. P. Stepanov, K. P. Aleshin, R. A. Manopov, B. N. Sannilov, V. V. Sklyarevsky, and V. G. Stankevich, Phys. Rev. Lett., 6, 155, (1963).
25. D. S. Bunbury, J. A. Elliott, H. E. Hall, and I. M. Williams, Phys. Rev. Lett., 6, 34, (1963).
26. P. P. Craig, and N. Sutin, Phys. Rev. Lett., 11, 460, (1963).
27. K. S. Singwi, and A. Sjölanoler, Phys. Rev., 120, 1093, (1960).
28. Yu. M. Kagan, Report to the All-Union Conference on Solid-State Physics, Moscow, December, 1963.
29. Yu. M. Kagan, Zhur. Eksp. i Teoret. Fiz., 46, (1964).

USE OF NUCLEAR QUADRUPOLE RESONANCE IN CHEMICAL CRYSTALLOGRAPHY

Reprinted from Journal of Structural Chemistry, Volume 1, Numbers 1 and 2

USE OF NUCLEAR QUADRUPOLE RESONANCE IN CHEMICAL CRYSTALLOGRAPHY

I. NATURE OF THE EFFECT

G. K. Semin and E. I. Fedin

Institute of Heteroorganic Compounds, Academy of Sciences of the USSR
Translated from Zhurnal Strukturnoi Khimii, Vol. 1, No. 2, pp. 252-267, July-August, 1960
Original paper submitted December 14, 1959

This review deals with applications to molecular structure and crystal structure. The first part is concerned with the main features of the effects responsible for nuclear quadrupole resonance spectra; relationships needed in practical work are presented. The second part gives a survey of the results so far obtained and brief descriptions of apparatus.

1. Energy Levels

Dehmelt and Kruger [1] made the first observations of nuclear quadrupole resonance (NQR) in 1949; it is one of the more recently discovered forms of magnetic resonance [2, 3]. It forms a division of spectroscopy and is concerned with the energy levels of solids; those levels have spacings corresponding to electromagnetic frequencies in the range from 1 to 1000 Mcps. First we must consider the nature of these levels.

Very often a nuclide has nuclei whose form differs from spherical; the electric quadrupole moment of the nucleus is a measure of the deviation from spherical form:

$$Q = \frac{1}{e} \int \rho \, (3z_m^2 - r^2) \, dx \, dy \, dz \tag{1}$$

in which e is the electronic charge, ρ is the density distribution of the charge in the nucleus, $r^2 = x^2 + y^2 + z^2$, and z_m is an axis coincident in direction with the nuclear spin vector I.

We see that Q is positive if the nucleus is elongated along the spin vector and is negative if the nucleus is shortened in that direction (Fig. 1).

Nuclear theory [4] shows that this moment is related to I: Q = 0 if I is 0 or $^1/_2$, and Q differs from zero only if $I \geq 1$. Table 1 gives the values at present known; those nuclides with M < 40 not appearing in Table 1 are known for certain to have Q = 0.

Suppose we have an atom having a finite moment present in a molecule forming part of a crystal. In Fig. 2 this atom is A; it is bound to atom A_1 by a covalent bond. Along the direction of this bond 0z there is a strong and inhomogeneous electric field specified by the field gradient $q_{zz} = \partial^2 V/\partial z^2$, in which V is the electrostatic potential produced by nearby charges. Strictly speaking, q is a tensor quantity having nine components in

TABLE 1

Atomic No.	Element	Mass No.	Natural abundance, %	Spin	Magnetic moment in magnetons		Gyromagnetic ratio	Quadrupole moment, 10^{-24} cm^2		Ratio of quadrupole moments
					[43]	[44]		[43]	[44]	
1	H(D)	2	0.015	1	+0.857348 ± 3	+0.857392	653.5	+0.00274	+0.002738	
3	Li	6	7.2	1	+0.82189 ± 4	+0.82193	626.7	+0.0005 ± 5	—	$Q_6/Q_7 = 1.9 \cdot 10^{-2}$
»	»	7	92.7	3/2	+3.2559 ± 1	+3.25598	1655.0	—0.012	—	
4	Be	9	100	3/2	—1.1774 ± 8	—1.1772	598.7	+0.02	+0.02	
5	B	10	18.83	3	+1.8004 ± 7	+1.8004	457.6	+0.06	+0.086	$Q_{10}/Q_{11} = 2.084$
»	»	11	81.17	3/2	+2.6886 ± 3	+2.68798	1367.0	+0.036	+0.042	
7	Na	14	99.62	1	+0,40365 ± 3	+0.4036	307.7	+0.02	+0.02	
8	O	17	0,4	5/2	—1.8929 ± 2	—1.89295	577.2	—0.004	—0.005	
11	Na	23	100.0	3/2	+2.2171 ± 2	+2.2161	1126.7	+0.11	+0.10	
13	Al	27	100.0	5/2	+3.6408 ± 4	+3.63853	1110.0	+0.156	+0.150	
16	S	33	0.74	3/2	+0.6429 ± 1	+0.6427	326.7	—0.065	—0.067	
17	Cl	35	75.4	3/2	+0.8219 ± 2	+0.82086	417.3	—0.0789	—0.085	
»	»	37	24.6	3/2	+0.6841 ± 2	+0.68330	347.4	—0.0621	—0,067	$Q_{35}/Q_{37} = 1.2688$
19	K	39	93.2	3/2	+0.391 ± 1	+0.39094	198.7	+0.14	—	$Q_{39}/Q_{41} = 1.220$
»	»	41	6.8	3/2	+0.215 ± 1	+0.21506	109.2	—	—	
21	Sc	45	100.0	7/2	+4,749			—0,22 *		
23	V	51	100,0	7/2	+5.1478 ± 5	+5.138	1121.0	—	+0.3	
25	Mn	55	100.0	5/2	+3.4677 ± 4	+3.4611	1056.0	0.5	+0.4	
27	Co	59	100.0	7/2	+4.6482 ± 6	+4.6389	1011,0	—	+0.5	
29	Cu	63	68.9	3/2	+2.2262 ± 4	+2.2213	1131.0	—0.16	—0.16	$Q_{63}/Q_{65} = 1.0806$
»	»	65	31.1	3/2	+2.3845 ± 4	+2,3790	1213.0	—0.15	—0.14	
31	Ga	69	60.1	3/2	+2.017 ± 1	+2,0108	1023.0	+0.18	+0.24	$Q_{69}/Q_{71} = 1.5867$
»	»	71	39.9	3/2	+2.561 ± 1	+2.5549	1299.0	+0.11	+0.15	
32	Ge	73	7.8	9/2	—0.8767 ± 1	—0.87680	148.8	—0.21	—0.22	
33	As	75	100.0	3/2	+1.4347 ± 3	+1.43491	729,3	+0.3	—0.3	
35	Br	79	50.6	3/2	+2,1058 ± 4	+2.0990	1070.0	+0.36	+0.33	$Q_{79}/Q_{81} = 1.19707$
»	»	81	49.4	3/2	+2.2696 ± 5	+2.2626	1153.0	+0.28	+0.28	
36	Kr	83	11.5	9/2	—0.9704	—0.966	164.0	+0.15	+0.16	
37	Rb	85	72.8	5/2	+1.3532 ± 4	+1.3482	411.3	+0.3	+2.8	$Q_{87}/Q_{85} = 2.07$
»	»	87	27.2	3/2	+2.7501 ± 5	+2.7414	1397.0	+0,14	+1.4	

*From [62].

TABLE 1 (continued)

Atomic No.	Element	Mass. No.	Natural abundance, %	Spin	Magnetic moment in magnetons		Gyromagnetic ratio	Quadrupole moment, 10^{-24} cm^2		Ratio of quadrupole moments
					[43]	[44]		[43]	[44]	
41	Nb	93	100,0	9/2	+6.166 ± 3	+6.1435	1045.0	—	—	
49	In	113	4,5	9/2	+5.486 ± 3	+5.4962	931.1	+0,75	+1.18	Q_{115}/Q_{113} = 1,0146
»	»	115	95.5	9/2	+5,500 ± 3	+5.5074	933.1	+0.76	+1,20	
51	Sb	121	57.25	5/2	+3.3591 ± 5	+3.3416	1019.0	—0,5 —1.2	—1.3	Q_{123}/Q_{125}=1.27475
»	»	123	42.75	7/2	+2.5465 ± 5	+2.5334	551.9	—0.7 —1.5	—1.8	
53	I	127	100.0	5/2	+2.8086 ± 8	+2,7938	856,5	—0,6 ± 2	—0.61	
54	Xe	131	21.18	3/2	+0.70	+0,704	410.0	—0,12	—0.12	
55	Cs	133	100.0	7/2	+2.5771 ± 9	+2.5642	561,7	—0,003	—0.03	
57	La	138	0,089	5	—2.776 ± 3	+3.68	605.0	—	+3.0	
»	»	139	100.0	7/2		+2,7615	—	+0,3	+0.9	Q_{138}/Q_{139}=3.0
59	Pr	141	100.0	5/2	+3.9	+3.8	1160.0	—0.05	—0,054	
60	Nd	143	12.2	7/2	—1.0 ± 2	—1.0	220.0	1.2	—	
»	»	145	8.3	7/2	—0.65 ± 9	—0.62	140.0	1.2	—	
62	Sm	147	17.0	7/2	—0.68 ± 8	—0.68	147.0	0.72	—	
»	»	149	15,0	7/2	—0.55 ± 6	—0,55	119.0	0,72	—	
63	Eu	151	47,8	5/2	+3.4	+3.4	1000.0	+1,2	+1,2	
»	»	153	52,2	5/2	+1.5	+1,5	460,0	+2,5	+2.6	
68	Er	167	22.94	7/2	—	—	—	—	+10.0	
70	Yb	173	17.02	5/2	—0.65	—0.66	198,0	+3.9	+0,4	
71	Lu	175	97.5	7/2	+2.9 ± 5	+2.9	570.0	+5.9	+6,5	
»	»	176	2.60	7	—	+4.2	—	—	+8.0	
73	Ta	181	100.0	7/2	+2.1	+2.1	460,0	+6.0	+7.0	
75	Re	185	31.7	5/2	+3,1433 ± 6	+3,1438	958.4	+2.8	+2.9	
»	»	187	62.9	5/2	+3.1755 ± 6	+3.1760	968.1	+2,6	+2.7	Q_{185}/Q_{187}=1.06
76	Os	189	16.1	3/2	+0,65	+0.65066	360,0	2.0	+2,0	
77	Ir	191	38.5	3/2	+0,17	+0.17	81.0	+1.5	+1,5	
»	»	193	61.5	3/2	+0,17	+0.17	86.0	+1.5	+1.5	
79	Au	197	100,0	3/2	+0,20	+0.14	69.1	+0.6	—	
80	Hg	201	13.18	3/2	—0,5590 ± 1	—0.607	308.0	+0.65	+0,5	
83	Bi	209	100,0	9/2	+4.0801 ± 5	+4,388	684,3	—0,4	—0,4	
92	U	235	0.715	(5/2)	—	—	—	—	—	

Cartesian coordinates x, y, z; an exact mathematical description of the interaction between the inhomogeneous field and the aspheric nucleus is extremely complicated. The usual approach is to assume that q is axially symmetric with respect to z, i.e., that $\partial^2 V/\partial x^2 = \partial^2 V/\partial y^2$, and to select axes such that the tensor for q takes a diagonal form. Then quantum mechanics [5] may be used to show that the energy arising from Q in the field of q takes the form

$$E_m = \frac{eq_{zz}Q\,[3m^2 - I\,(I+1)]}{4I\,(2I-1)} \tag{2}$$

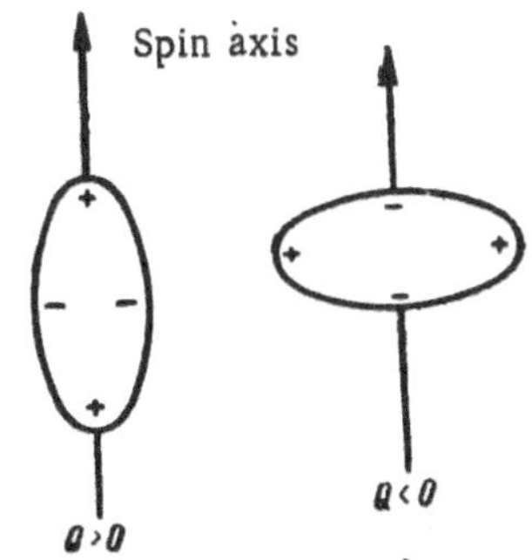

Fig. 1. Quadrupole configurations.

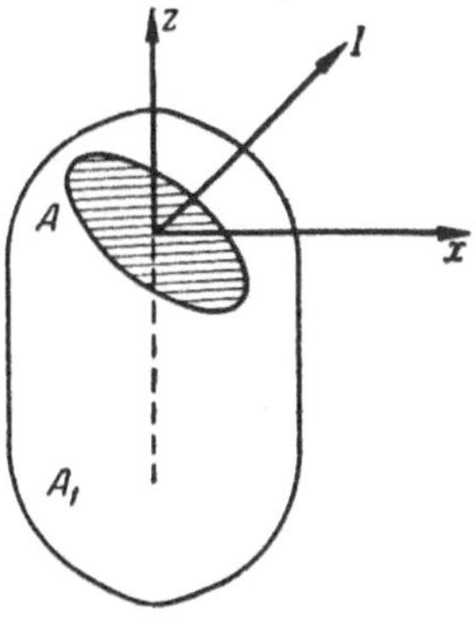

Fig. 2. Asymmetric nucleus in the inhomogeneous electric field of a molecule; z is the bond direction.

Here m is the magnetic quantum number, which takes 2I + 1 values between I and -I. We see from (2) that the energy levels are doubly degenerate with respect to m, because E_m depends only on $|m|$. For example, Table 1 and (2) show that N and Cl have only two energy levels each; transitions between these levels can produce only one emission or absorption line. These transitions can be stimulated by an electromagnetic field whose quanta have the energy

$$h\nu = E_{(m)} - E_{(m-1)}. \tag{3}$$

The Boltzmann distribution implies that the $E_{(m-1)}$ level will be the more densely populated; the difference in the population densities will be $E_{(m)}$; i.e., it will be extremely small at room temperature. This difference is responsible for the absorption; the tendency for a field in accordance with (3) to produce transitions upward will be greater than the tendency for the upper level to produce emission. Although the energy levels of (2) are electrical in origin, the transitions are caused by the interaction between the magnetic component of the field and the magnetic moment of the nucleus. The absorption line is usually very narrow, $\Delta\nu$ being such that $\nu/\Delta\nu \geq 10^4$. (The factors that influence the line width are dealt with below.)

From (2) and (3) we have the resonance frequency as

$$\nu_r = \frac{3eQq_{zz}}{4hI\,(2I-1)}(2|m_I|-1), \tag{4}$$

in which m_I is the larger of the two magnetic quantum numbers for the transition. The quantity eQq/h is usually called the quadrupole coupling constant and is expressed in Mcps (Planck's constant is omitted in writing the quantity). Then we have $\nu_r = 3eQq/4$ for I = 1, eQq/2 for I = $^3/_2$, and so on.

Very often q is not axially symmetric; then we use the asymmetry parameter $\eta = (q_{xx} - q_{yy})/q_{zz}$. In this case the formula for the energy levels is a function of I and is more complicated than (2). Table 2 gives equations that have been derived [5-7].

Formulas have been derived [2] for the resonance frequencies from the equations of Table 2; in terms of η they are:

TABLE 2

I	Equation	Units of E
3/2	$E^2 - 3\eta^2 - 9 = 0$	$A = \frac{eQq_{zz}}{4I(2I-1)}$
5/2	$E^3 - 7(3+\eta^2)E^2 - 20(1-\eta^2) = 0$	$2A$
7/2	$E^4 - 42\left(1+\frac{\eta^3}{3}\right)E^2 - 64(1-\eta^2)E + +105\left(1+\frac{\eta^2}{3}\right)^2 = 0$	$3A$
9/2	$E^5 - 11(3+\eta^2)E^3 - 44(1-\eta^2)E^2 + +\frac{44}{3}(3+\eta^2)^2E + 48(3+\eta^3)(1-\eta^2) = 0$	$6A$

$$I = 1;\ \nu = \frac{3eQ\,(q_{zz})}{4}\cdot\left(1 \pm \frac{\eta}{3}\right)$$

$$I = {}^3/_2;\ \nu = \frac{eQ\,|\,q_{zz}\,|}{2}\cdot\left(1+\frac{\eta^2}{3}\right)^{\frac{1}{2}}$$

$$I = {}^5/_2;\ \nu_1 = \frac{3eQ\,|\,q_{zz}\,|}{20}\cdot(1 + 1.09259)\,\eta^2 - 0.63403\,\eta^4)$$

$$\nu_2 = \frac{3eQ\,|\,q_{zz}\,|}{10}\cdot(1 - 0.20370\,\eta^2 + 0.16215\,\eta^4)$$

$$I = {}^7/_2;\ \nu_1 = \frac{eQ\,|\,q_{zz}\,|}{14}\cdot(1 + 3.63333\,\eta^2 - 7.26070\,\eta^4) \tag{5}$$

$$\nu_2 = \frac{2eQ\,|\,q_{zz}\,|}{14}\cdot(1 - 0.56667\,\eta^2 + 1.85952\,\eta^4)$$

$$\nu_3 = \frac{3eQ\,|\,q_{zz}\,|}{14}\cdot(1 - 0.1000\,\eta^2 - 0.01804\,\eta^4)$$

$$I = {}^9/_2;\ \nu_1 = \frac{eQ\,|\,q_{zz}\,|}{24}\cdot(1 + 9.03333\,\eta^2 - 45.69070\,\eta^4)$$

$$\nu_2 = \frac{2eQ\,|\,q_{zz}\,|}{24}\cdot(1 - 1.338095\,\eta^2 + 11.72240\,\eta^4)$$

$$\nu_3 = \frac{3eQ\,|\,q_{zz}\,|}{24}\cdot(1 - 0.18571\,\eta^2 - 0.12329\,\eta^4)$$

$$\nu_4 = \frac{4eQ\,|\,q_{zz}\,|}{24}\cdot(1 - 0.08095\,\eta^2 - 0.004258\,\eta^4)$$

We see that the spectrum becomes more complicated as the spin increases. We cannot determine eQq and η from the resonance frequency of an unknown compound if I = $^3/_2$, but these quantities can be deduced (from the Zeeman effect) if a monocrystal of that compound is available. In all other cases the number of lines is sufficient to establish these quantities. We shall see below that both of those quantities give us valuable information about the molecule. Cohen [7] gives a table of energy levels for $0 \leq \eta \leq 1$.

Experiment gives us eQq_{zz} and η, which may appear at first sight very abstract quantities of little value to those concerned with molecular structure. However, even the early experiments demonstrated that eQq_{zz} and η are very sensitive to details of the structures of the molecule and crystal. We shall see in Section 7 that q_{zz} is very sensitive to the structure of the electron shells and that eQq_{zz} gives us unique information about the degree of ionic character and (sometimes) the hybridization of covalent bonds. In many cases η enables us to estimate the π-electron concentration near the atom.

The most important feature is that these effects on eQq_{zz} and η are not small. For example, the frequencies for Cl^{35} in trichloromethane and dichloromethane respectively are 38.2809 and 35.9912 Mcps, i.e., they are differentiated in the second significant figure, whereas the frequencies themselves can be measured to six or seven significant figures. The values of η are found to vary from 0.05 to 0.8.

Nonequivalent positions in the unit cell give rise to well-resolved distinct lines. For example, $CHCl_3$ gives resonances at 38.3081 and 38.2537 Mcps, which implies that the crystal contains molecules in two distinct sets of positions. These splittings exceed the line widths by one to two orders of magnitude. Phase transitions and other effects can be detected.

The results obtained in this way are dealt with systematically in the second part of this review.

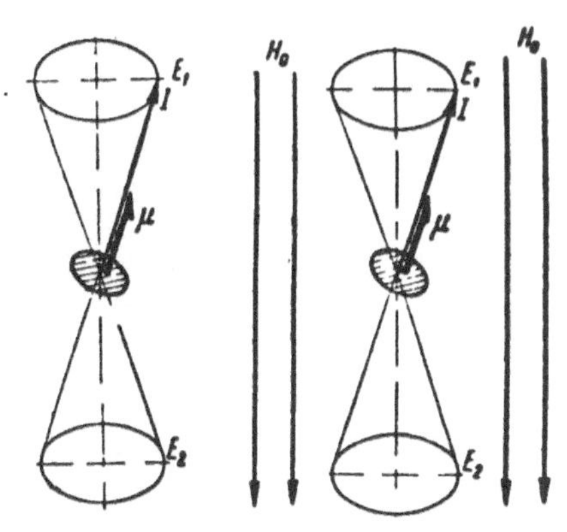

Fig. 3. Nuclear precession in the absence and in the presence of an external constant magnetic field H_0. On the left $E_1 = E_2$, on the right $E_1 > E_2$.

2. Zeeman Effect in NQR

The above description of quadrupole resonance may be illustrated by considering the precession of the nuclear spin about the bond direction 0z. Then the various E_m correspond to various angles between I and the axis of precession (Fig. 2). This model illustrates the degeneracy implied by (2), because, for a given m, the energy of the state in which I makes an angle θ with the axis does not differ from the energy for the state in which the angle is $180° - \theta$.

However, an external fixed magnetic field H_0 produces a change in the picture; if H_0 is oriented as in Fig. 3, the angle θ corresponds to an energy higher than the energy corresponding to $180° - \theta$. Now the energy difference $\Delta E \ll kT$ for these two positions of $\vec{\mu}_I$, so the two levels will be populated and the degeneracy with respect to m has been removed. Thus the NQR line is split into two components by the external field (Fig. 3).

The ultimate effect of this splitting takes one form for a monocrystal and another for a polycrystalline material; in the first case the signals from all nuclei combine to give a resolved spectrum, whereas the random arrangement in a polycrystalline material merely causes the line to broaden, with the result that the peak strength of the signal is reduced. This behavior is used to identify the signal; a NQR line is broadened by a weak magnetic field, so the absence of broadening indicates that the line has another origin.

In the simplest case the field H_0 produces a splitting equal to [7]

$$\Delta E = \pm m\mu_I H_0 \cos\theta / I, \tag{6}$$

in which θ is the angle between H_0 and the axis of q, and μ_I/I is the gyromagnetic ratio (417.214 cps/gauss for Cl^{35}). We see that the splitting is zero if H_0 lies at 90° to the axis of q. A more complicated case is that in which H_0 removes the degeneracy of an $E_{1/2}$ level; the states with $m = +1/2$ and $m = -1/2$ have $\Delta m = 1$ and both interact with H_0. Then H_0 produces mixed levels described by

$$E_{\pm} = \frac{eQq_{zz}}{4I(2I-1)}\left[\frac{3}{4} - I(I+1)\right] \pm \left[\cos^2\theta + \left(I + \frac{1}{2}\right)^2 \sin^2\theta\right] \frac{\mu_I}{2_I} H_0. \tag{7}$$

Now (7) implies that zero splitting cannot occur for any θ. Figure 4 shows that the splitting produces four observed lines when $I = 3/2$; the $E_+ \rightleftharpoons E_-$ transitions are not observed at ordinary H_0, because the corresponding frequency is less than ν_I by some two orders of magnitude. Figure 4, with (6) and (7), shows that the internal lines have a frequency separation

$$\nu_{\alpha\alpha'} = 3\frac{\mu_I}{I} H_0 \cos\theta + \frac{\mu_I}{I} H_0 \sqrt{\cos^2\theta + \left(I + \frac{1}{2}\right)^2 \sin^2\theta}. \tag{8}$$

Again, (8) implies that these two lines (the strongest) become one lying at ν_I for a certain θ; for $I = 3/2$ this angle is given by (8) as $\tan\theta = \sqrt{2}$, i.e., $\theta = 54°44'$. This feature is used in structure studies on monocrystals; the crystal is rotated to find the directions for H_0 that give no splitting, and those directions form a cone whose axis is the direction of the bond.

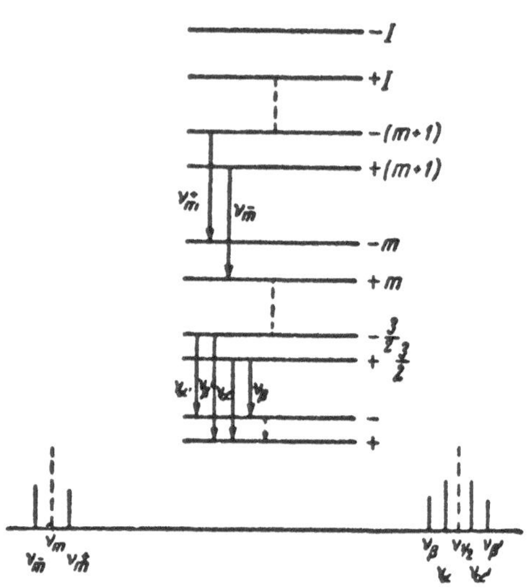

Fig. 4. Zeeman splitting of quadrupole levels.

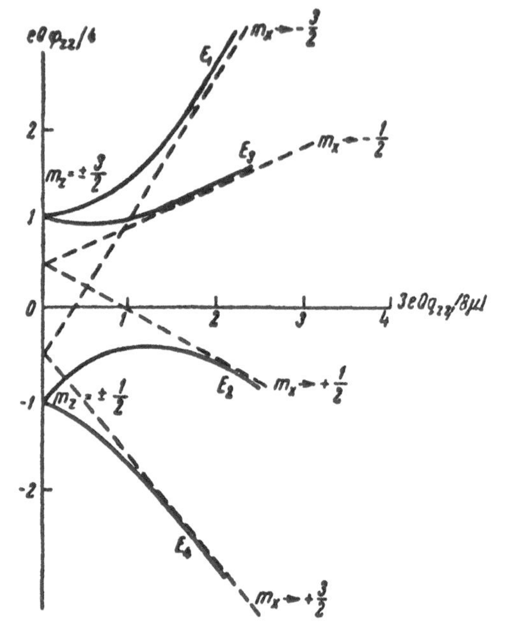

Fig. 5. Nonlinear variation of splitting with H_0 in strong fields.

Formulas (6), (7), and (8) are correct for small H_0. Figure 5 (from [10]) shows results for $I = ^3/_2, \eta = 0$, and $\theta = 90°$. Here the abscissa is H_0 in relative units. This orientation of the field is such that the energy levels can be calculated [11-13] by means of a second approximation in perturbation theory. We see that the splitting varies very nonlinearly with the field as $H_0 \rightarrow 3eQq_{zz}/8\mu_I$. This field is $H_0 = \frac{3 \cdot 108 \cdot 10^6 \cdot 6.6 \cdot 10^{-27}}{8 \cdot 0.82 \cdot 0.5 \cdot 10^{-23}} = 65{,}400$ gauss for Cl^{35}; by strong fields are meant those with $H_0 = (5\text{-}15) \cdot 10^3$ gauss in practical work, so formulas (6), (7), and (8) are quite adequate in all cases.

3. Line Width

The difference in population density corresponding to (2) is extremely small, but the resonance is very often detectable at or above 300°K, so we must conclude that $\Delta\nu$ is very small ($\nu_r/\Delta\nu \geq 10^5$) for an isolated molecule. If the width were unchanged in the crystal, it would always be possible (though perhaps troublesome) to detect the resonance, and the sensitivity would not need to be any higher than the value commonly found in NMR spectrometers, which have been commercially available for some years.

However, things are not always so favorable; the line is often much broadened, with the result that the peak height is reduced (so that the sensitivity must be much higher). The problem of line width is that of detecting the resonance at all; it requires the most careful attention. The main factors affecting the line width are dealt with in the next section.

4. Static Factors Affecting the Line Width

Magnetic interactions are the most important; the nuclei surrounding the species of interest may have magnetic moments as well, so the local magnetic fields have to be considered. These fields produce an internal Zeeman effect, which may appear either as a splitting or (more often) as a mere broadening (the strength and directions of the fields are decisive here). We may differentiate between direct magnetic dipole-dipole interactions and indirect interactions via the electrons in the bonds (the I bond effect).

The dipole-dipole interaction occurs between immediately adjacent nuclei, in which case a fine structure is seen, as in the case of HIO_3 [14]. Here the effects of the fields of more distant nuclei are suppressed by the field H of the nucleus and cause only a certain broadening of the fine-structure components.

Alternatively, no dipole-dipole interaction may be possible between nearest neighbors, in which case the other nuclei having dipole moments produce only broadening. Koi et al. [15] have discussed the natural line widths (those controlled by the lifetimes of the quadrupole states) and the broadening produced by dipole-dipole interactions. If the line width found for some elements is that caused by the lifetime, then the ratio of the line widths for two isotopes of that element equals the ratio of the squares of the quadrupole moments. This principle has been extended [16, 17] in careful theoretical and experimental studies on $NaClO_3$ and $NaBrO_3$. Exact line-width calculations based on the work of Abragam et al. [18] have shown that dipole-dipole interactions are responsible for the major part of the line width in each case, although the actual width is not very much larger than the natural width for $NaBrO_3$.

The indirect interaction can affect the line shape only via interactions confined within the molecule; the splitting is appreciable for Br_2, I_2, and ICl [19, 20]. Kozima et al. [21] have discussed the splitting found for Br_2; theory predicts a spectrum with 18 lines (9 produced by $Br^{79}Br^{79}$ and $Br^{81}Br^{81}$, and 9 others produced by $Br^{79}Br^{81}$). The width of each line exceeds 10 kcps, so the spectrum cannot be resolved; experiment reveals a broad line on which can be seen three weak peaks, the shape of the line corresponding roughly with the calculated one. The interaction constant J is about 3 kcps.

It is stated [14, 15] that terrestrial fields can affect the line width in some cases; Koi has observed that the earth's field broadens the very narrow line of $NaBrO_3$ (the line becomes narrower when this field is eliminated). Livingston and Zeldes report a similar effect for $KClO_3$, whose line is exceptionally narrow (width about 290 cps).

The third static factor is associated with defects and stresses in the crystal. Defects cause a random variation in q, and so the line is broadened. Stresses (e.g., produced by rapid growth, etc.) have the same effect, as Allen and Dean [22, 23, 26] were the first to observe. Wang [24] has found that the lines of Sb and Cl in $SbCl_3$ become broader when the temperature is reduced; he explains the effect in terms of stresses induced in the crystal. Fuke and Koi [16] have confirmed the effect, though they find that Wang's values for the line widths are wrong. We also find that those [24] results for the line width of Sb at room temperature are too small. Koi et al. [25] have compared the measured line width for $KBrO_3$ with the one calculated on the basis of dipole-dipole interactions; they concluded that imperfections in the crystal account for a great deal of the broadening. (The systematic work of Duchesne and Monfils on the effect of impurities is dealt with below.) Das and Hahn [5] point out that an exact theory of this effect cannot be produced unless we know the law of the molecular interaction and the effects of that interaction on the electric field gradient. This interesting problem has not yet been solved.

5. Dynamic Factors Influencing the Line Width

The motion of the molecule in the crystal can affect the line width; thermal oscillations restrict the lifetime and control the natural line width. The effects are much smaller than those of the static factors, although such oscillations may also have effects resembling those of defects: they may cause an extra spread in q and so broaden the line.

The spin-lattice relaxation time T_1 is used to estimate these effects. Quanta of energy satisfying (3) act on the set of levels and make the population densities on the states equal.

When the radiofrequency field is turned off, the difference in population density Δn produced by the field tends to return to the original difference Δn_0 (which is determined by the temperature) in accordance with the law

$$\Delta n = \Delta n_0 (1 - e^{-t/T_1}), \tag{9}$$

in which t is time elapsed.

The excited state has a long lifetime if T_1 is large; in this case the thermal processes may be unable to maintain any appreciable difference in population density, with the result that, after a certain time, the number of quanta absorbed in unit time equals the number emitted in that time, and the specimen ceases to produce a signal. This power-saturation effect is avoided in NMR by keeping the radiofrequency power as small as possible. It is very much more difficult to produce saturation in quadrupole resonance. (Gutowsky et al. [27] give T_1 for p-dichlorobenzene as 20 msec, for example.) All the same, although the signal may not vanish, the line width may be increased. See Wang [24] for details.

The spin-spin relaxation time T_2^* is used to specify the line shape more directly. Let the function $g(\nu)$ describe the distribution of resonance frequencies produced by all factors acting together. This function is normalized via the condition $\int_{-\infty}^{\infty} g(\nu)d\nu = 1$. Then we have

$$T_2^* = 2g(\nu_r). \quad (10)$$

The line shape gives us information about the relaxation processes. Spin-echo methods (see part II) provide the most accurate means of measuring T_1 and T_2^*. Stresses in the crystal affect the line width via the spin-spin relaxation, i.e., via T_2^*. The value of T_2^* for p-dichlorobenzene varies from 0.5 msec to 70 μsec, although T_1 is not affected [27].

Livingston [28] has observed in BCl_3 an interesting effect of the molecular vibrations. The line for Cl is a doublet showing a small splitting (2.8 kcps), although the crystal structure is such as to exclude the possibility that there are two positions for the molecule in the unit cell; again, the dipole-dipole interactions are too weak to explain the effect. Douglas [5] explains the effect in terms of the internal vibrations of the molecule; the B^{10} vibrates more rapidly, so the resonance frequency for $B^{10}Cl_3$ should be lower (see section 6). The $B^{10}Cl_3$ and $B^{11}Cl_3$ molecules differ in the motion of their B atoms. This difference is found to give an effect on q_{zz} equivalent to 2 kcps. The rest of the splitting is caused by dipole-dipole interactions. Douglas's view is confirmed by the fact that it predicts the intensity ratio of the doublet correctly. Now B^{10} is the less abundant isotope, so the low-frequency component of the doublet should be the weaker one, as is actually the case. The effect is best observed on molecules in which D replaces H. Duchesne et al. [29] have used CH_3Br^{79}, CD_3Br^{79}, and so on for this purpose, but they found that the large splittings are the reverse of those predicted on the basis of Douglas's explanation. Douglas [30] has calculated the structure of the resonance line of Cl_2, which Crane et al. [31] have examined. The results are used to explain the isotope effects of Cl^{35} and Cl^{37} on q_{zz} in terms of rotary oscillations about the center of gravity of the molecule.

Several studies have been made of the effects of inhibiting molecular rotation [32-34].

6. Dynamic Effects That Affect the Resonance Frequency

Bayer [35] has considered the effects of rotary oscillations on q_{zz}, i.e., on the frequency; Raman and infrared spectra indicate that the lowest frequencies for such oscillations are at least two orders of magnitude higher than the highest possible frequencies for quadrupole resonance. Bayer deduces from this the important result that the nucleus is in an electric field whose q differs from that for a molecule at rest (the effective q is a function of temperature). His calculation for $I = 3/2$ gives the resonance frequency as a function of temperature in the form

$$\frac{1}{\nu_r}\cdot\frac{d\nu}{dT} = -\frac{3h^2}{8\pi^2kT^2}\left[\frac{e^{h\nu_x/kT}}{A_x\,(e^{h\nu_x/kT}-1)^2} + \frac{e^{h\nu_y/kT}}{A_y\,(e^{h\nu_y/kT}-1)^2}\right] \quad (11)$$

in which ν_r is the frequency for $\eta = 0$, ν' is the same corrected for η, T is temperature, ν_x and ν_y are the frequencies of the oscillations about the x and y axes of the tensor for q, and A_x and A_y are the corresponding moments of inertia. Oscillations on z do not contribute to q (they affect only η), and (5), with $I = 3/2$, may be used to demonstrate that this effect may be neglected in (11). Figure 6 shows a curve calculated from (11); we see that the frequency rises as the temperature falls. The temperature coefficient for p-dibromobenzene is 18.7 kcps/deg; that for p-dichlorobenzene is 2.7 kcps/deg. These values are such that the specimen must be kept at a fixed temperature; even a difference of 2-3°C between parts of a specimen can broaden the line very considerably.

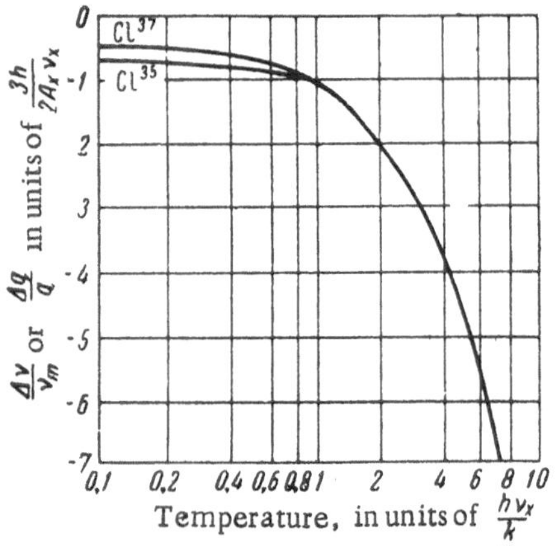

Fig. 6. Resonance frequencies for Cl^{35} and Cl^{37} as functions of temperature according to Bayer's theory.

Bayer's predicted result is found to agree with the results for most compounds, although anomalous behavior is not unknown. Barnes and Engardt [36] have found that $TiBr_4$ gives a frequency that increases between room temperature and -50°C but thereafter decreases. This anomaly cannot be explained on Bayer's theory, but Kushida et al. [37] and Gutowsky et al. [38] have interpreted it by means of a theory in which allowance is made for the pressure produced within the lattice when the temperature is lowered. If

$$\left|-\frac{\alpha}{\varkappa}\left(\frac{\partial\nu}{\partial P}\right)_T\right| > \left|\left(\frac{\partial\nu}{\partial T}\right)_V\right|$$

(in which α is the thermal expansion coefficient and κ is compressibility), then the temperature coefficient is positive; the Bayer mechanism is then a minor one, as occurs at low temperatures, for Fig. 6 shows that $(\partial\nu/\partial T)_V$ falls as $T \to 0$. This can occur only if $(\partial\nu/\partial P)_T < 0$. The sign of this derivative has been examined for $TiBr_4$ [36]; it actually is negative, so any detailed consideration of the frequency as a function of temperature must take account of dynamic and static factors.

7. The Quadrupole Interaction Constant

No exact theory is available to relate q to the electron-density distribution. Townes and Dailey's semi-empirical formulas [5, 42, 44] are nearly always used in interpreting results, although they give only a rough general estimate of the bond type. However, the terminology they introduced has become so widely used that we have adhered to it in what follows.

We follow Das and Hahn [5] and consider Townes and Dailey's theory in relation to ICl. The external electrons have the configuration $5s^25p^5$ in the I atom and $3s^23p^5$ in the Cl atom; i.e., one p electron is needed to complete the shell. A closed shell would have spherical symmetry, so the electric field gradient at the nucleus would be zero. Therefore Townes and Dailey assume that the actual values of q in the free atoms are caused by lack of a 5p or 3p electron respectively; i.e., they equal the q caused by one p electron, apart from sign. The electron orbitals overlap when ICl is formed; for simplicity it is assumed that the main overlap occurs in the p_z orbital forming a σ bond. (The possible effects of π bonds formed from p_x and p_y orbitals are considered in Section 8.)

The electrons forming the σ bond have an orbital ψ that is a linear combination [47, 49] of the atomic orbitals ψ_{Cl} and ψ_I:

$$\psi = a\psi_{Cl} + b\psi_I \tag{12}$$

Here a and b are normalized in accordance with the condition $a^2 + b^2 + 2abS = 1$, in which $S = \int \psi_{Cl}\psi_I d\tau$ is the overlap integral. We have a = b for a purely covalent bond, as occurs in Cl_2, I_2, etc., but $a \neq b$ if the atoms are different. Pauling [54] states that in the latter case the bond is partly ionic, a measure of the proportion of ionic character being

$$i = |a^2 - b^2|. \tag{13}$$

The bond is fully ionic if b = 0; in this case we would have Cl^- (with a closed shell) and I^+ (lacking two p electrons). This does not in fact occur, and the I-Cl bond is intermediate in type; the p_z orbital of Cl contains $1 + i$ electrons, and that of I contains $1 - i$ electrons.

Then (12) may be used to find the field gradient at the center of the Cl atom. By definition we have

$$q_{zz} = \frac{\partial^2}{\partial z^2}\left(\frac{e}{r}\right) = \frac{e\,(3z^2 - r^2)}{r^5} = \frac{e\,(3\cos^2\theta - 1)}{r^3}, \tag{14}$$

in which r is the distance from the charge e to the nucleus, and θ is the angle between r and the z axis. The exact q_{zz} is to be found by averaging (14) with respect to (12):

$$\begin{aligned} q_{zz} &= e\int (a\psi_{Cl} + b\psi_I)\frac{3\cos^2\theta - 1}{r^3}(a\psi_{Cl}^* + b\psi_I^*)\,d\tau = \\ &= a^2e\int \psi_{Cl}\frac{3\cos^2\theta - 1}{r^3}\psi_{Cl}^*\,d\tau + 2abe\int \psi_{Cl}\frac{3\cos^2\theta - 1}{r^3}\psi_I^*\,d\tau + \\ &+ b^2e\int \psi_I\frac{3\cos^2\theta - 1}{r^3}\psi_I^*\,d\tau. \end{aligned} \tag{15}$$

The first term on the right in (15) is a^2 times the gradient q_0 produced by one electron in the orbital ψ_{Cl} so that the magnetic quantum number of that electron is $m_l = 0$; the second is the gradient produced by the electron density in the overlap region, and the third is the contribution from ψ_I. Now the gradient varies as $1/r^3$, so we can neglect the third term, and even the second at the expense of some loss of accuracy. Then

$$q_{zz} \approx a^2q_0. \tag{16}$$

From (13) and the normalization condition we have

$$a^2 = \frac{1+i}{2(1+S)}. \tag{17}$$

Experiment and general arguments show that S can be neglected in (17); although the overlap governs the bond energy, it has little effect on the nuclear interaction. Then

$$2q_{zz} = (1+i)\, q_0. \tag{18}$$

In addition to the two σ electrons, which make a contribution to q_{zz}, there are many other electrons that may make a certain contribution (to $q_{zz\ Cl}$, for example) in ICl. Those electrons are: a) two pairs of $3p_x$ and $3p_y$ electrons (π electrons) of Cl, b) the $3s^2$ electrons, c) the inner ($1s^2 2s^2 2p^6$) electrons of Cl, and d) the joint action of all the charges in I, which supplies one 5p electron to the σ bond (this last Townes and Dailey allow for by means of one elementary positive charge placed at the center of the nucleus). Now q is calculated for the nucleus, which contains no electrons, so Laplace's equation applies, i.e,

$$q_{zz} + q_{xx} + q_{yy} = 0. \tag{19}$$

Now $q_{xx} = q_{yy}$ on account of the axial symmetry, so that

$$q_{p_x} = -\frac{1}{2} q_0, \tag{20}$$

and each π electron produces a gradient

$$q_{p_x} = -\frac{1}{2} q_{p_z} = q_{p_y} \tag{21}$$

while all the π electrons of the Cl atom together produce a gradient at the center of the Cl nucleus of

$$q_\pi = -2q_0. \tag{22}$$

The contributions under b) and c) above are smaller, but they ought to be incorporated, strictly speaking. Townes and Dailey neglected them.

The contribution to q under d) is of the order of $1/r^3$ (r is the distance between the nuclei); this correction is usually negligible.

Thus the p electrons are found to produce the main contribution to q. Let N_x, N_y, N_z denote the populations of the $3p_x$, $3p_y$, and $3p_z$ states of the Cl atom; then (21) and (22) imply that the p electrons make a contribution to q_{zz} of

$$q_{zz} = \left(\frac{N_x + N_y}{2} - N_z\right) \cdot q_0 = U_p q_0, \tag{23}$$

in which q_0 is as defined above for (16) for the p_z electron of Cl.

Here U_p is the number of unbalanced p electrons; its sign is taken as positive when there is a deficit of p electrons along the z axis.

Free atoms are encountered in molecular beams and show their effects in atomic spectra; here the projection of the orbital moment on the special direction is maximal ($m_l = \pm 1$). Experiments of this kind give the standard data for quadrupole interactions in atoms, so it would be best to derive q_{zz} in terms of such states. Then (21) and (23) give us that

$$eQq_{zz} = eQU_p q_0 = -2\, eQq_{\pm 1} U_p = -2U_p (eQq)_{\text{atom}}. \tag{24}$$

Then, with (18) on the assumption that the σ bond is ionic, and with $N_x = N_y = 2$ in (23), we have

$$eQq_{zz} = -(1-i)(eQq)_{\text{atom}}. \tag{25}$$

In Section 1 we stated that NQR experiments give us the quadrupole interaction constant for a molecule in a crystal. Now (25) shows that the eQq_{zz} found in this way can be used to calculate the proportion of ionic character. For ICl we have $eQq_{zz} = -82.50$ Mcps [47], and $(eQq)_{\text{atom}} = 109.74$ Mcps. Then (25) gives $i = 0.25$, a value that agrees with Gordy's view [45, 47] that sp hybridization need not be considered in relation to the bonding orbitals of Cl (in which he differs from Townes and Dailey). The long discussion of this topic [45-49] has not yet produced any clear-cut results. Let us consider how sp hybridization for I-Cl would affect the above results. Let the degree of s character of the above bonding orbital of Cl be s. Here the number of effective p_z electrons in this orbital is reduced from (1 + i) to (1 + i)(1 − s), but the s orbital has now become a hybrid sp orbital of degree of hybridization s, so that it contains 2s effective p_z electrons. The total number of p_z electrons is than (1 + i) (1 − s) + 2s, so that a change occurs in the number of unbalanced p electrons:

$$U_p = 2 - ((1 + i)(1 - s) + 2s] = (1 - i)(1 - s). \qquad (26)$$

We see that (26) shows that sp hybridization produces the same effect as ionic character (it reduces the quadrupole interaction). Experiment gives only $|eQq_{zz}|$, so it is very difficult to choose between one form (in which all of the reduction in eQq is caused by the ionic character) and the other (in which hybridization makes a contribution as well); the problem can be resolved only by resorting to information obtained in other ways.

Townes and Dailey differ from Gordy in not concluding that the above eQq_{zz} and $(eQq)_{atom}$ for I-Cl imply that the I-Cl bond is rather ionic. They use Pauling's value i = 0.08 for this bond, and use (26) to find that the p_z orbital has 18% s hybridization. They survey a large number of bonds to formulate their hybridization rule: "If a halogen atom is bound to an atom B such that the difference of the electronegativities is $(x_A - x_B) \geq 0.25$, then the A-B bond has s = 0.15 for its degree of s character; s hybridization is absent if $(x_A - x_B) < 0.25$."

Figure 7 illustrates the difference between the two approaches (ionic character as a function of the difference in the electronegativities). Gordy's curve predicts a greater degree of ionic character, but both run far above the curves compiled earlier on the basis of dipole moments. Quite apart from the difficulty about the exact proportion of s hybridization, it is clear that the quadrupole interaction results imply that the bonds formed by the halogens are much more ionic than had been supposed previously [47].

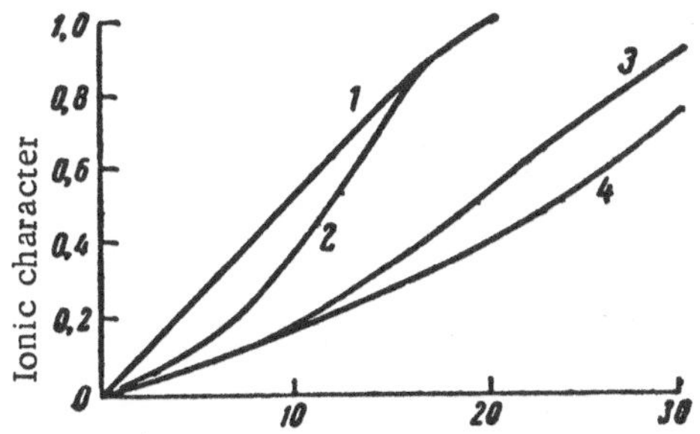

Fig. 7. Degree of ionic character of a bond A-B as a function of $x_A - x_B$. 1) Gordy; 2) Townes and Dailey; 3) Pauling (from dipole moments); 4) Hannay and Smith [53] (from dipole moments).

An interesting feature of Fig. 7 is that the quadrupole results predict that the bond will be 100% ionic if $(x_A - x_B) \geq 2$. Wilmhurst [50] has proposed a relation

$$i = \frac{|x_A - x_B|}{x_A + x_B}. \qquad (27)$$

which implies that universal curves $i = f(x_A - x_B)$ cannot be constructed; all we can do is to construct curves relating i to x_B or $x_A - x_B$ for each kind of atom A. Wilmhurst has used the quadrupole results to estimate the s character of carbon-halogen bonds; his values of s come out much larger even than Townes and Dailey's, which means that the i given in [47] are to be reduced correspondingly.

Further work is required to establish the true position and to resolve this conflict of views.

We must point out that Townes and Dailey's semiempirical treatment can give no rigorous basis for choosing between the various ways of separating the contributions to eQq from ionic character and hybridization (the more so as regards determining the type of hybridization). Experiment gives only |eQq|, and results on hybridization and ionic character so obtained will remain somewhat undefined while we lack a proper theory of quadrupole interactions. Both difficulties can be obviated if we proceed purely empirically by measuring the eQq for extensive series of related compounds; on this basis conclusions can be drawn about the nature of the bonds from the trend in the values.

More rigorous treatments have been attempted; Bassompiere [55-57] has performed an exceptionally laborious rigorous calculation of the eQq for nitrogen in HCN and has obtained satisfactory agreement with experiment. The volume of calculation is such that no similar calculation has yet been attempted for a more complicated molecule. Benderskii and Blyumenfel'd [58, 59] have found that the contribution to the $(eQq)_{atom}$ from polarization of the filled shells by the unbalanced electron may be as much as 20-30% for elements in the fourth and subsequent groups. (This implies that we are justified in neglecting polarization for Cl but not for I in the example above.) They have also proposed a new method of computing the wave functions for diatomic molecules in terms of measured eQq and dipole moments; in this way the degree of ionic character and the s character can be estimated directly without resort to further assumptions. Calculations for HCl and HBr show that the method is usable, but serious difficulties are encountered in applying it to more complex molecules.

Schatz [60, 61] has adopted the entirely opposite approach (that of calculating eQq from the parameters of the molecular orbital), but his results cannot be compared with experiment, for these parameters are unknown.

We conclude that there is at present no sound theoretical basis for calculating the parameters of bonds from measured quadrupole interaction constants. The importance of NQR to chemistry is not that it enables one to estimate these parameters but that eQq is exceptionally sensitive to small changes in the character of a bond. In the second part we give some examples which demonstrate that the changes in eQq are additive for chloroalkanes, that eQq is proportional to Hamett's constant σ, etc. Much of value is to be expected from empirical laws of this kind derived from NQR studies.

8. Asymmetry in the Field Gradient

Formulas (25) and (26) relate to the case $\eta = 0$, although this is not so in most cases; then the interaction constant varies not only with the bond type but also on account of the fact that $N_x + N_y$ in (23) deviates from the value of 4 proper to a σ bond. But the change in the p_x and p_y orbitals is equivalent to the production of a π bond, so two bond types are involved.

Let us consider ICl again; the degree to which the second type of bond is present is π, in which π may be defined as the difference between p_x (or p_y) for the free Cl (or I) atom and the same quantity for the bound Cl (or I) atom.

Then the contribution to q_{zz} from the π electrons is reduced by πq_{atom} relative to the contribution for the case in which only a single bond type is involved [this case corresponds to (22)]. Then (26) is replaced by

$$U_d = (1 - i)(1 - s) - \pi. \tag{28}$$

Now we consider the relation of π to η, for which purpose we introduce in (23) the symbol $(U_p)_z$ for the number of unbalanced p electrons; then

$$(U_p)_x = N_x - \frac{N_y + N_z}{2}, \qquad (U_p)_y = N_y - \frac{N_x + N_z}{2}. \tag{29}$$

Hence $q_{xx} = (U_p)_x \cdot q_0$, $q_{yy} = (U_p)_y \cdot q_0$, so the asymmetry parameter is

$$\eta = \frac{q_{xx} - q_{yy}}{q_{zz}} = \frac{\frac{3}{2}(N_x - N_y)}{N_z - \frac{N_x + N_y}{2}}. \tag{30}$$

If η is not too large, we can neglect η^2 above and in (5), in which case we have

$$I = {}^3/_2; \quad \eta = \frac{3(N_x - N_y)(eQq)_{atom}}{4\nu}, \qquad I = {}^5/_2; \quad \eta = \frac{3(N_x - N_y)(eQq)_{atom}}{2\nu}. \tag{31}$$

In (31) we have ν as the frequency of the $E_{1/2} \rightleftharpoons E_{3/2}$ transition, so that the asymmetry parameter is a direct measure of the difference in the populations of the $3p_x$ and $3p_y$ states. This difference is a direct measure of the degree of duality in the bond for a planar molecule; recent studies [51, 52] of substituted benzenes and heterocyclic compounds have given some interesting results, which we shall consider in the second part of this review.

There are some compounds whose molecules are axially symmetric about a double bond, in which case duality in the bond type causes no appreciable difference in the $3p_x$ and $3p_y$ orbitals [5]; the asymmetry parameter is zero. It is more difficult to examine the bond type for such a molecule by means of NQR.

LITERATURE CITED

1. H. Dehmelt and H. Krüger, Naturwissenschaften 37, 11 (1950).
2. D. J. E. Ingram, Spectroscopy at Radio and Microwave Frequencies [Russian translation] (IL, Moscow, 1959).
3. E. R. Andrew, Nuclear Magnetic Resonance [Russian translation] (IL, Moscow, 1958).
4. V. V. Malyarov, Theory of the Nucleus [in Russian] (Fizmatgiz, Moscow, 1959).
5. T. P. Das and E. L. Hahn, Nuclear Quadrupole Resonance Spectrocopy (New York, 1958).
6. R. Bersohn, J. Chem. Phys. 20, 1505 (1952).
7. M. Cohen, Phys. Rev. 96, 5, 1278 (1954).
8. R. Livingston, J. Chem. Phys. 26, 1102 (1957).
9. H. Meal, J. Chem. Phys. 24, 1011 (1956).
10. H. Krüger, Z. Phys. 130, 371 (1951).
11. Whitmer, Weidner, Hsiang and Weiss, Phys. Rev. 74, 1478 (1948).
12. R. V. Pound, Phys. Rev. 79, 685 (1950).
13. G. Becker and H. Krüger, Naturwissenschaften 38, 121 (1951).
14. R. Livingston and H. Zeldes, J. Chem. Phys. 26, 351 (1957).
15. V. Koi, A. Tsujimura and T. Fuke, J. Chem. Phys. 23, 1346 (1955).
16. T. Fuke and V. Koi, J. Chem. Phys. 29, 973 (1958).
17. K. Kano, J. Phys. Soc. Japan 13, 975 (1958).
18. A. Abragam and K. Kambe, Phys. Rev. 91, 1184 (1953).
19. S. Kojima and K. Tsukada, J. Phys. Soc. Japan 10, 591 (1955).
20. G. D. Watkins and R. Walker, Bull. Amer. Phys. Soc. (II) 1, 11 (1956).
21. S. Kojima, et al., J. Phys. Soc. Japan 12, 1225 (1957).
22. H. Allen, J. Amer. Chem. Soc. 74, 6074 (1952).
23. C. Dean, J. Chem. Phys. 23, 1734 (1955).
24. T. C. Wang, Phys. Rev. 99, 566 (1955).
25. V. Koi, A. Tsujimura and V. Imaeda, J. Chem. Phys. 27, 603 (1957).
26. C. Dean, Phys. Rev. 96, 1053 (1954).
27. H. Gutowsky and D. Woessner, J. Chem. Phys. 27, 1072 (1957).
28. R. Livingston, J. Phys. Chem. 57, 496 (1953).
29. J. Duchesne, A. Monfils and J. Garson, Physica 22, 817 (1956).
30. D. C. Douglas, J. Chem. Phys. 28, 504 (1959).
31. L. Crane, R. Anderson and H. Robinson, Bull. Am. Phys. Soc. (II) 4, 11 (1959).
32. M. Buyle-Bodin, Ann. Phys. 10, 533 (1955).
33. H. Dodgen and I. Ragle, J. Chem. Phys. 25, 376 (1956).
34. E. Lassetre and C. Dean, J. Chem. Phys. 17, 31 (1949).
35. H. Bayer, Z. Phys. 130, 227 (1951).
36. R. G. Barnes and R. Engardt, J. Chem. Phys. 28, 731 (1958).
37. T. Kushida, G. Benedeck and N. Bloemberger, Phys. Rev. 104, 1364 (1956).
38. H. Gutowsky and G. Williams, Phys. Rev. 105, 464 (1957).
39. R. Barnes and W. Smith, Phys. Rev. 93, 95 (1954).
40. R. Bacher and S. Goudsmith, Atomic Energy States (New York, 1932); C. E. Moore, Atomic Energy Levels, Nat. Bur. Standards. Circ. 467 (Washington, 1949).
41. H. Bethe and R. Bacher, Rev. Mod. Phys. 8, 226 (1936).
42. C. Townes and B. Dailey, J. Chem. Phys. 17, 782 (1949).
43. A. Lösche, Kerninduction (Berlin, 1957).
44. C. Townes and A. Schawlow, Microwave Spectroscopy (New York, 1955).
45. W. Gordy, J. Chem. Phys. 19, 792 (1951).
46. W. Gordy, Disc. Faraday Soc., 18 (1955).
47. Spectroscopy in Chemistry [in Russian] (IL, Moscow, 1959), Chap. II.
48. B. Dailey, Disc. Faraday Soc., 18 (1955).

49. W. J. Orville-Thomas, Quart. Rev., XI, 162 (1957).
50. J. Wilmshurst, J. Chem. Phys. 30, 561 (1959).
51. F. Adrian, J. Chem. Phys. 29, 1381 (1958).
52. M. Dewar and E. Lucken, J. Chem. Soc. (London), 2653 (1958).
53. N. Hannay and C. Smith, J. Amer. Chem. Soc. 68, 171 (1946).
54. L. Pauling, Nature of the Chemical Bond [Russian translation] (Goskhimizdat, Moscow, 1947).
55. L. Bassompiere, Disc. Faraday Soc., 18 (1955).
56. L. Bassompiere, J. Chem. Phys. 21, 614 (1954).
57. L. Bassompiere, Compt. rend. Acad. Sci. 240, 285 (1955).
58. V. A. Benderskii and L. A. Blyumenfel'd, Optika i Spektroskopiya 3, 402 (1957).
59. V. A. Benderskii, Diploma Thesis [in Russian] (Izd. MGU, 1959).
60. P. Schatz, J. Chem. Phys. 21, 730 (1954).
61. P. Schatz, J. Chem. Phys. 22, 695 (1954).
62. G. Fricke, et al., Phys. Verhandl. 10, 13 (1959).

USE OF NUCLEAR QUADRUPOLE RESONANCE IN CHEMICAL CRYSTALLOGRAPHY

II. METHODS AND RESULTS OF EXPERIMENTAL INVESTIGATIONS

É. I. Fedin and G. K. Semin

Institute of Heteroorganic Compounds, Academy of Sciences of the USSR
Translated from Zhurnal Strukturnoi Khimii, Vol. 1, No. 4, pp. 464-499, November-December, 1960
Original paper submitted August 25, 1960

A brief description of quadrupole radiospectrometers is given. The main trends in work on nuclear quadrupole resonance (NQR) are defined: attempts to investigate electron density distribution in molecules; searches for empirical relationships between chemical properties and NQR frequencies; applications of NQR to certain problems of the structure of real crystals. Data on the NQR spectra of 477 different chemical compounds have been collected for the first time.

1. Frequency Ranges

The frequency at which quadrupole resonance occurs depends (see Part I of this review) on the absolute values of the quadrupole moment Q of the nucleus, the gradient q of the electric field created in the center of the nucleus by all the neighboring charges, and the spin function $f(I)$. The table of the quadrupole moments of the stable isotopes, given in Part I, clearly shows that values of Q vary periodically with increase of mass number. The "normal" tendency is an increase of Q down each group of the periodic system. Exceptions to this tendency are Cs^{133} in group I, and Bi^{209} in group V.

The value of q is determined by many factors, and may vary severalfold for a given atom in accordance with the state of the chemical bond.

The nuclear spin I also varies from nucleus to nucleus, resulting in appreciable variations of $f(I)$.

The consequence of the combined influence of these three factors is that, for example, chlorine exhibits NQR in the range from 5 to 70 Mcps, and bromine in the range from 40 to 270 Mcps.

Figure 1 shows the frequency ranges in which transitions between quadrupole energy levels of various elements in different chemical compounds have been observed. It can be seen that a quadrupole radiospectrometer intended for observation of the resonance of all isotopes suitable for the purpose should cover a very wide range of frequencies, from the long-wave radiofrequency range to decimeter waves.

2. Apparatus for Observation of NQR

It should be emphasized that continuous searching for an unknown weak signal in a wide frequency range is an extremely difficult problem which cannot as yet be regarded as finally solved. The apparatus described below was used by most workers, not because entirely satisfactory solutions to the problem had been found, but owing to absence of better equipment.

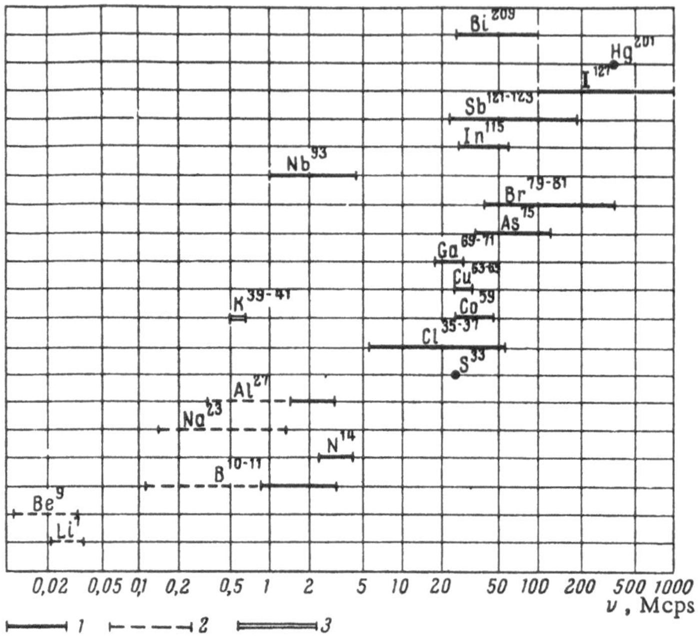

Fig. 1. Chemical shifts of NQR frequencies for different elements, measured: 1) by means of NQR; 2) from splitting of magnetic resonance lines; 3) by the pulse technique.

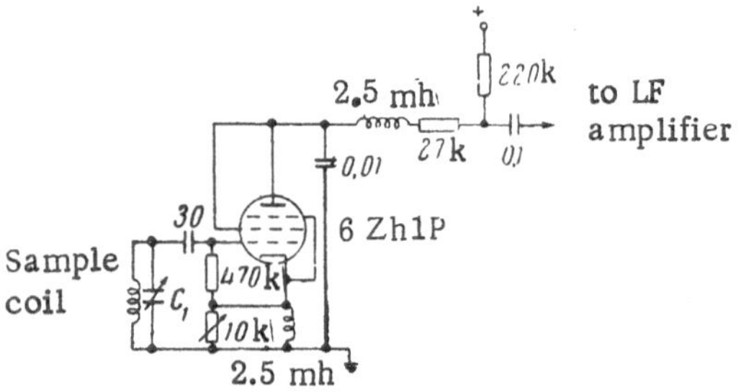

Fig. 2. The Hopkins autodyne circuit.

The techniques are based on the use of slightly modified autodyne circuits of the type developed for observations of nuclear magnetic resonance [1]. In most NQR spectrometers, the signal is produced by a regenerative or superregenerative oscillating detector. The processes in the regenerative oscillating detector may be briefly described as follows.

The substance to be investigated is placed in the coil of the oscillator circuit (Fig. 2). Oscillations, the frequency of which can be varied smoothly (for example, by means of a variable capacitor), are induced in the circuit by suitable selection of feedback. When the frequency ν of the oscillations in the circuit satisfies the condition $h\nu = \Delta E$, where ΔE is the distance between the quadrupole levels, the substance in the coil begins to absorb energy from the resonant circuit, varying the active conductance component of the latter. Whereas, before resonance, the conductance of the circuit is G, at the instant of resonance it is $(1 + \gamma)G$. Calculations show that changes in the conductance of the circuit result in changes of the high-frequency amplitude; an amplitude of U_0 before resonance becomes $(1 + \epsilon)U_0$ at resonance. The modulation of the circuit conductance γ is proportional to the imaginary component χ' of the magnetic susceptibility of the specimen, the filling factor η, and, with certain assumptions [1], the figure of merit of the circuit Q_0:

$$\gamma = -4\pi\eta\chi'' Q_0. \tag{1}$$

The filling factor η is the ratio of the high-frequency energy in the specimen to the total energy of the circuit.

The imaginary component of magnetic susceptibility [2] is determined by the properties of the specimen

$$\chi'' = f(I)\cdot\frac{N_0\mu^2}{kT}\cdot\frac{\nu}{\Delta\nu}, \tag{2}$$

where N_0 is the number of resonating nuclei in 1 cm^3; μ is their magnetic moment; ν is the frequency of the NQR line; $\Delta\nu$ is its half-width. The spin function $f(I) \sim 1$. It follows from Eqs. (1) and (2) that, for the strongest

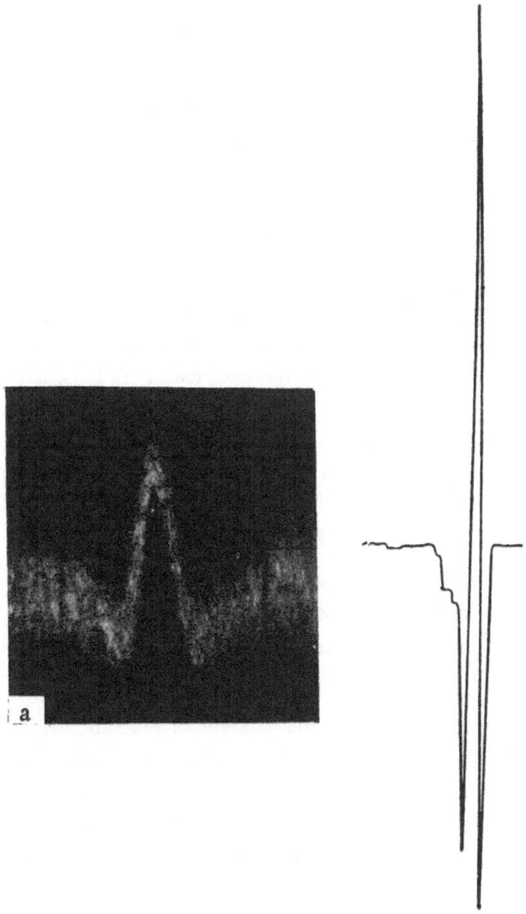

b

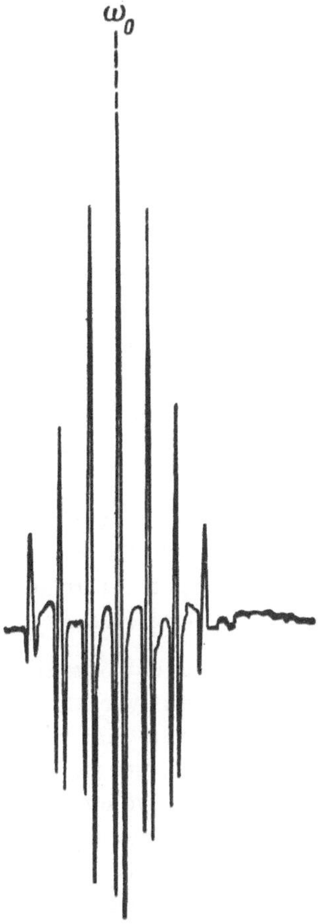

Fig. 3. a) NQR signal from 20 cm^3 of sodium chlorate, observed by means of a regenerator and broad-band amplifier; b) same signal after synchronous detection at frequency $2\nu_m$.

Fig. 4. Frequency trace in superregenerator

NQR signals at $Q_0 \sim 10^2$, the conductance modulation has a very low value: $\gamma \lesssim 10^{-4}$. Spectrometer design calculations must be based on values of γ of the order of 10^{-6} to 10^{-7}.

The question of the relative change ϵ of the circuit voltage at a given γ was considered by Shpigel' et al. [3] and by Buyle-Bodin [4]. In [3], the transition of the oscillator from one steady state to another on introduction of additional quenching into the circuit is considered. It was shown that the change of the circuit voltage is directly proportional to γ and inversely proportional to the voltage before introduction of the additional quenching. The relationship $\epsilon \sim \gamma U_0^{-1}$ prevents the use of high radio-frequency voltages in the circuit, and usually $U_0 \leq 1$ v in NQR observations with the aid of regenerative spectrometers. Approximate calculations based on the data in [3] show that for the usual NQR signals, $\epsilon \sim 10\gamma \sim (10^{-5}\text{-}10^{-6})$. In these calculations, only the operation of the autodyne as an oscillator was taken into account, and detection conditions were not considered. Buyle-Bodin's investigation was concerned with detection in the regenerative oscillating detector: the ϵ/γ ratio and the signal form were shown to depend on the time constant of the detecting circuit. It follows from calculations and from Buyle-Bodin's experiments that at maximum sensitivity, which is the basic goal of nearly all workers in spectrometer tuning, the signal reproduceu by the detector is very far from the true form of the NQR absorption line. Examination of theoretical work and experimental results leads to the conclusion that the possibilities of raising the sensitivity of regenerative oscillating detectors are almost exhausted. Many regenerative NQR spectrometers are described in the literature. They are all either different versions of the Hopkins circuit [5-9], or modifications of the Pound circuit [1, 10-15], and are all of much the same sensitivity. The Pound circuit is preferable, because it embodies a successful solution of the problem of stabilization of the generator and detector with large frequency changes.

Wang's attempt [8] to raise the sensitivity of the Hopkins circuit by introduction of feedback from the detected NQR signal led to an interesting version of the regenerator. However, G. K. Semin's measurements showed that, in the cases when Wang's circuit gives a real sensitivity increase, it introduces considerable distortion into the signal form.

Very small voltage changes in the circuit are detected by the modulation method: the frequency ν_0 is swung over a narrow range in accordance with the equation $\nu = \nu_0 + \Delta\nu_m \sin 2\pi \nu_m t$. The frequency deviation amplitude $\Delta\nu_m$ chosen is approximately 5-10 times the width of the NQR line; the deviation frequency is chosen in the 20-300 cps range. Sometimes magnetic modulation is used instead of frequency modulation: a magnetic field sufficient for complete blurring of the signal is applied at frequency ν_m to the coil (see Part I, Section 2). In either case, if $\nu_0 = \nu_r$, the voltage at the generator tube grid is modulated by the audio frequency ν_m with a modulation amplitude equal to the signal amplitude ϵU_0. As the grid current–grid voltage relationship is nonlinear, and tube detects this voltage, and pulsation at frequency ν_m appears in the anode current. This pulse potential is fed to a low-frequency amplifier (LFA) with a transmission band of the order of 1-5 kcps, sufficient to let through, without serious distortion, the harmonics necessary for reproduction of the signal form.

The amplified signal voltage is fed to the vertical plates of an electronic oscillograph, the scanning of which is synchronized by the modulating voltage. A signal similar to that shown in Fig. 3a then appears on the screen. It can be seen that with this "wide-band" method of observation the signal is comparable with fluctuation noise of the generator-detector.

A fairly obvious way of raising the signal-to-noise ratio is by increasing the amount of substance taken for the investigation. In view of the weakness of the NQR signals, it is important that the specimen should occupy a volume in which most of the energy of the radiofrequency magnetic field is concentrated, i.e., such that $\eta = 1$. Livingston [6] proposed the use of metallic containers for the sample coils for this purpose. Usually, with a coil about 1.5-5 cm^3 in volume, from 20 to 70 cm^3 of a polycrystalline specimen is placed in the container. Naturally, different regions of the specimen in the container are connected in different ways with the coil and make unequal contributions to the observed signal. Ė. I. Fedin's measurements of field strength in a container showed that with this technique a tenfold increase of specimen volume gives roughly a twofold increase of signal-to-noise ratio.

Unfortunately, such large amounts of substance are not usually available. It is then necessary to decrease η and to use ampoules inserted into the coil. The usual volume of our ampoules is 0.5-2 cm^3. The use of ampoules has one important advantage: if necessary, the specimen can be sealed, and after determination of the quadrupole spectrum it can be recovered without any damage for further investigation. However, with ampoules of this size, the NQR spectra of very many compounds cannot be seen on the oscilloscope screen with broad-band amplification. Thus, the sensitivity of regenerative quadrupole spectrometers very often proves inadequate. Therefore, superregenerative oscillating detectors, giving five- to tenfold sensitivity gains, are widely used in work on nuclear quadrupole resonance. In a superregenerator the radiofrequency oscillations are not continuous but are "quenched" at a quenching frequency ν_q satisfying the condition $\nu_m \ll \nu_q \ll \nu_0$. The sensitivity gain is achieved by formation of the signal under nonsteady generation conditions and by the effective use of very high radiofrequency voltages (up to 40 v) in the circuit. A characteristic feature of the superregenerator is the multicomponent character of its resonance curve: in addition to oscillations at the circuit tuning frequency ν_0, oscillations of frequencies $\nu_0 \pm n\nu_q$, where n = 1, 2, 3, ..., exist in the circuit. Therefore, in passing through a single quadrupole resonance line, the superregenerative spectrometer records 2n + 1 NQR "signals," of which the central one is the true signal, and the others are superregenerative side-band responses. This multiplication of the signals often hinders accurate measurements of NQR frequencies and analysis of NQR multiplet spectra. However, this same property of the superregenerator often proves useful for identification of weak NQR signals which may be confused with accidental distortion of the zero line. The reason is that, first, the superregenerative side-band responses of the quadrupole signal vary regularly in amplitude and are separated by equal frequency intervals; and second, that they alternate in phase. Figure 4 shows the result of a measurement of the frequency of a superregenerative oscillating detector. The oscillating detector operated as a highly sensitive radio receiver, receiving on a small antenna the almost monochromatic signal emitted by a standard-signal oscillator. As a result, a multicomponent frequency trace was recorded with slow variation of the frequency ν_0 of the oscillating

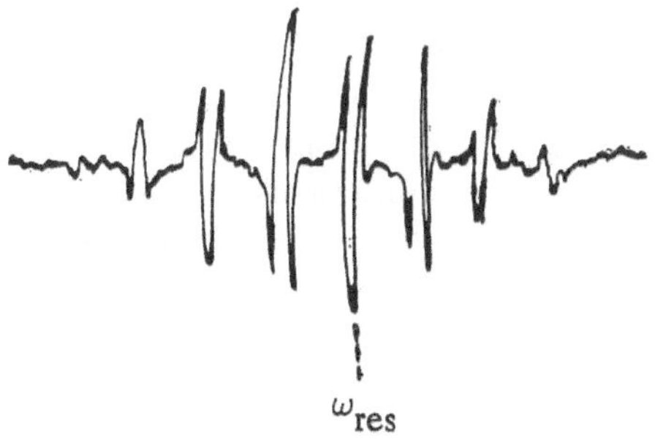

Fig. 5. NQR signal of the rare isotope Cl^{37} from 1 g of p-dichlorobenzene, obtained with a superregenerative spectrometer.

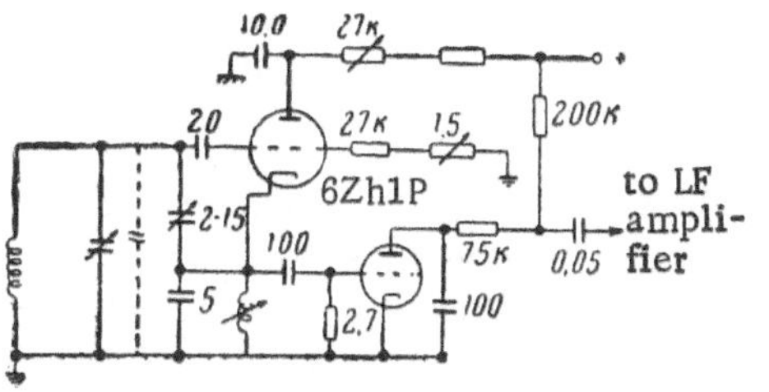

Fig. 6. Superregenerative oscillating detector for the chlorine band.

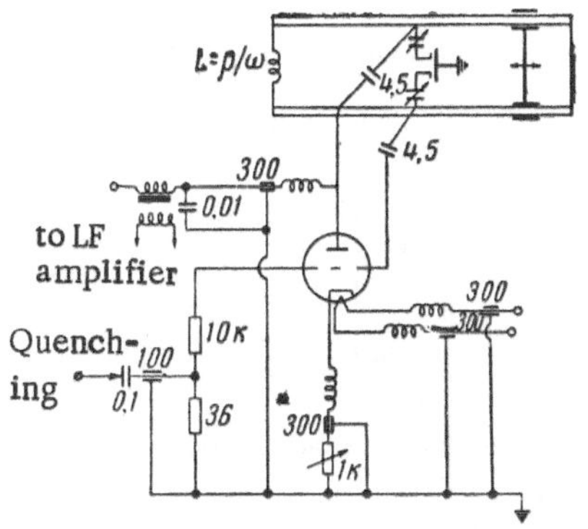

Fig. 7. Superregenerative oscillating detector for the bromine band.

detector. It can be seen that the peaks of the superregenerative spectrum of this trace are all of the same polarity. Figure 5 shows a trace of an NQR line. Here, alternating polarity of the superregenerative components of the signal is clearly seen.

Our measurements suggest that the superregenerative oscillating detector reacts to a greater extent to radiation of energy from a nuclear spin system than to the corresponding absorption of energy.

Dean and Pollak [16] described an ingenious method for eliminating the superregenerative side-band response of the signal by slow modulation of the quenching frequency ν_q. Variations of ν_q cause the components to shift to the right and left of the undisplaced position, so that in synchronous detection (see below) these components cannot accumulate in the RC circuit and become "blurred." The circuit tuning frequency ν_0 does not depend on ν_q, and the true NQR signal is recorded almost without loss of intensity.

The physical processes taking place during formation of the NQR signal in the superregenerator are very complex. It can be regarded as experimentally proven that, because of this, the true form of the quadrupole resonance line undergoes considerable distortion in observations with the superregenerative spectrometer [17]. This has to be accepted in cases where high sensitivity is required and precise measurement of the NQR line widths is not needed.

Our superregenerative oscillating detector circuits are shown in Figs. 6 and 7. The oscillating detector for the chlorine band (Fig. 6) has a circuit with lumped constants. It is assembled in accordance with Dean's circuit [18], with the difference that a separate tube is used for detection of the NQR signal. This is to ensure reliable operation of the detector during changes of the generator operating conditions in searches for unknown resonance frequencies. A similar device was used independently by Safin [19], and also by Buyle-Bodin [4] with a regenerator.

At frequencies from 100 to 350 Mcps, satisfactory results were obtained with the two-wire circuit shown in Fig. 7. Push-pull oscillators give good results in this band [20-23]. Oscillators with coaxial resonators are used at higher frequencies [24].

Unfortunately, in most cases the sensitivity of the superregenerative spectrometer is still inadequate in wide-band signal observations.

For further increase of the spectrometer sensitivity, the transmission band of the amplifier must be narrowed sharply, as for a band width of $\Delta\nu$ the noise voltage at the amplifier output is proportional to $\sqrt{\Delta\nu}$ [1]. The synchronous detector [25] provides an effective means for narrowing the transmission band. If the synchronous detector is followed by an RC circuit with a time constant of 10 sec, the detector transmission band is equal to 0.1 cps. This means that, in comparison with broad-band observation of the signal, when $\Delta\nu \geq 500$ cps, the sensitivity can be raised 100-fold and more by synchronous detection, owing to narrowing of the transmission band. This is attained at the cost of a corresponding increase in the time required for the experiment. The signal from the synchronous detector is usually fed to a direct-current recording millivoltmeter. A typical trace of an NQR signal demonstrating the sensitivity gain achieved in this way is shown in Fig. 3b. The characteristic form of the signal on the trace is explained by the fact that the synchronous detector was tuned to the second harmonic of the modulation frequency. It is then possible to avoid the influence of parasitic amplitude modulation [26], which is a serious obstacle in NQR observations with a frequency-modulated regenerative spectrometer. Recording of the second harmonic of the signal was proposed by Livingston and examined in detail by Fedin and Konstantinov [26]. The need to tune the synchronous detector to frequency $2\nu_m$ is completely eliminated with magnetic modulation of the signal, and in some instances also with frequency modulation of superregenerative oscillating detectors, which are less liable to parasitic amplitude modulation than regenerators.

Examination of the signal in Fig. 3b may give a somewhat exaggerated idea of the sensitivity of modern quadrupole radiospectrometers. In reality, it is exceptionally difficult to detect a signal $\frac{1}{100}$ as strong as that shown in Fig. 3b if the frequency of this weak signal is not known. The reason is that, in prolonged frequency searching, the tuning of the oscillating detector is disturbed and zero stability is lost. It is not much easier to raise the zero stability than to improve the sensitivity of the oscillating detector.

The range of possible applications of nuclear quadrupole resonance could be greatly extended if signal generators of essentially less internal noise than the existing oscillating detectors could be devised. It is possible that success in this direction might be achieved by circuits with parametric excitation of oscillations, similar to Konstantinov's circuit [27].

An interesting variation of experiments in the NQR field was provided by work on quadrupole spin echo [28, 29]. If two rectangular radio-frequency pulses separated by time τ are applied to the specimen, then 2τ seconds after the first pulse an echo signal is produced in the specimen, caused by interaction of the damping signals of nuclear induction from the first and second pulses. Damping of the echo with known τ makes it possible to determine the time T_2 of spin–spin relaxation, and the relationship between the echo amplitude and the time τ between the pulses makes it possible to determine the spin–lattice relaxation time T_1 [17]. Searches for unknown signals by this technique have so far proved unsuccessful.

3. Experiments on Zeeman Splitting of NQR Spectra

Substances, single crystals of which have been studied by the NQR method with application of a weak constant magnetic field, are listed below:

Br^{81} (solid)	[21]	$NaClO_3$	[39]
I^{127} (solid)	[30]	$KClO_3$	[39]
p-$Cl_2C_6H_4$	[31]	$Ba(ClO_3)_2 \cdot H_2O$	[39]
p-$Br_2C_6H_4$	[32]	BBr_3	[40]
p-$ClC_6H_4NH_2$	[11, 33]	AsI_3	[41]
p-$ClC_6H_4CH_2Cl$	[34]	$AsI_3 \cdot 3S$	[41]
$(ClCN)_3$	[35–37]	HIO_3	[42]
1,2,4,5-$Cl_4C_6H_2$	[38]	Cu_2O*	

The most systematic investigation of Zeeman splitting of NQR spectra was that carried out by Dean et al. [38] in the case of 1,2,4,5-tetrachlorobenzene. Specially designed semiautomatic equipment was used, whereby it was possible to investigate fully the crystal structure of the material. The authors concluded that quadrupole resonance may prove of great help in x-ray structural investigations of certain substances. The accuracy in determinations of bond directions proved to be somewhat better than by the x-ray method; the error did not exceed $\pm 1°$.

Most investigations of Zeeman splitting of NQR signals for chlorine are undertaken for determination of the asymmetry parameter η, because (see Part I of this review), without experimental determination of η, data provided by quadrupole spectra on the nature of chemical bonds are too indefinite.

An interesting example of the use of values of η found by quadrupole resonance is provided by the work of Casabella and Bernes [43]. They succeeded in correlating the values of field gradient asymmetry parameter η_M in the center of the metal atom nucleus, measured for $AlBr_3$ and InI_3, with the valence bond angles formed by the metal atom with halogen atoms in dimeric molecules. The values of bond angles in $AlBr_3$ and InI_3 found from quadrupole resonance data were in quite satisfactory agreement with the values of the valence angles determined from x-ray diffraction data. This also apparently confirmed the bond hybridization model which forms the basis for interpretation of the values of η_M. Unfortunately, these authors apparently did not consider the

* Measured by I. A. Safin.

degree of certainty to which a particular hybridization model is determined in such experiments. However, the calculation idea is itself very valuable, because it is only by correlation of data on nuclear quadrupole interaction with data obtained by independent methods that any reliable information on electron distribution in the molecules of substances can be obtained. In general, it is probably preferable not to predetermine the form of the wave function, but to derive it to a certain degree of reliability in accordance with the experimental parameters (see Part I, Section 7).

4. Chemical Applications of NQR Spectra

It is evident from the example just given that, in many cases, charge distributions in molecules can be investigated in adequate detail by means of quadrupole resonance. Dewar and Lucken [44] point out that the information provided by NQR spectra on charge distribution is often unique, as it cannot be obtained by any other investigation method known at the present time.

These workers [44] investigated π bonds in chloro derivatives of nitrogen heterocycles. They also studied the character of C–Cl bonds in chloro derivatives of maleic anhydride, thiophene, and aniline salts [45]. In particular, they showed that the NQR frequencies of 2-chlorothiophene, 2,5-dichlorothiophene, and 2,3,4,5-tetrachlorothiophene are considerably higher than those of the corresponding chlorobenzenes (see table of NQR spectra). Dewar and Lucken concluded that the electronegativities of S and C cannot be assumed equal, as is usually done. They suggested that the effective electronegativity of the S atom in chlorothiophenes increases by transfer of one 3p electron to the 3d orbit. The C–Cl bond should then become less ionic in character; i.e., the NQR frequency of chlorine should increase; this is found to be the case. This example is an apt illustration of the range of ideas used in interpretation of chemical shifts of quadrupole resonance frequencies. The full table of NQR frequencies in various compounds, which is given in this review, offers the reader a wide field for such correlations and conclusions.

The linear relationship, discovered by Meal [46], between quadrupole frequencies in certain compounds and values of Hammett's constant σ is of considerable interest with regard to chemical applications of quadrupole resonance. It will be remembered that the semiempirical equation derived by Hammett in 1935 [47] is of the form

$$\ln \frac{k}{k_0} = \sigma\rho, \tag{3}$$

where k is the rate constant of a reaction proceeding in the position meta or para to a given substituent in the benzene ring; k_0 is the rate constant for the same reaction in unsubstituted benzene; σ and ρ are constants, the first of which characterizes the influence of the substituent, and the second is general for all substituted benzenes and is characteristic of the reaction. It was pointed out a long time ago [47] that σ is a measure of electron density changes at the site of the reaction, due to the substituent, while ρ measures the "sensitivity" of a given reaction to variations of electron density.

Meal [46] found that the quadrupole resonance frequencies of chlorine in various substituted chlorobenzenes vary linearly with the Hammett constant σ for a given position.

The results of Meal's investigations are shown in Fig. 8. It can be seen that the linear relationship between NQR frequency and σ is only very roughly approximate. It seemed likely that deviations from this relationship should prove no less important for chemical applications of NQR than the general relationship postulated by Meal. Bray's group, which works very productively in the NQR field, has devoted its efforts to attempts to interpret these deviations; a similar trend is characteristic of the work of Dewar and Lucken.

In a large series of investigations by Bray et al. [48-53], the approximate linear relationship between Hammett's constant σ and quadrupole resonance frequencies was confirmed for a very large number of substituted benzenes. The following formula for observed NQR frequencies was derived for various chlorobenzenes:

$$\nu_{obs} = (34.826 + 1.024\sigma) \text{ Mcps} \tag{4}$$

and for bromobenzenes

$$\nu\,(Br^{81})_{obs} = (227.02 + 7.866\sigma) \text{ Mcps.} \tag{5}$$

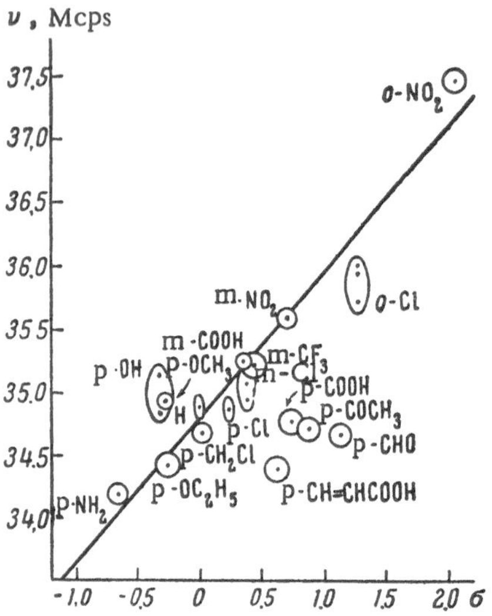

Fig. 8. Relationship between NQR frequencies in various chlorobenzenes and the Hammett constant σ.

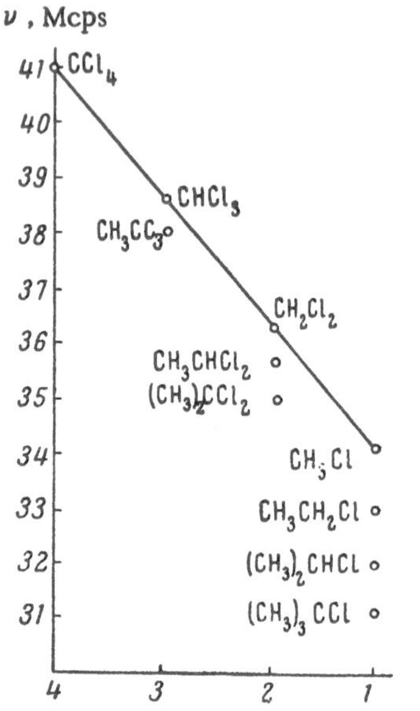

Fig. 9. NQR frequencies of chlorine in certain chloroalkanes.

The importance of these expressions cannot be overestimated: on the one hand, they show that it is possible to predict reactivities of various compounds from their NQR frequencies and, on the other, they offer a deeper insight into the relationship between charge distribution in a molecule and reactivity.

Bray et al. [51] attempted to apply Eqs. (1) and (2) to nitrogenous heterocycles, and attributed the observed deviations from linearity to increase of the double-bond character owing to introduction of nitrogen into the benzene ring. If this course is followed consecutively, large deviations between theory and experiment are found. These deviations may be caused either by the unjustified use for heterocycles of the values of σ obtained in experiments with substituted benzenes, or by the fact that the assumption of additivity of values of σ is invalid.

It must be emphasized that the failure of any particular model "explaining" the influence of different chemical environments on the NQR frequency does not justify pessimistic conclusions on the potentialities of the method. The most important fact is that the influence, which cannot always be explained as yet, exists, and is a powerful effect which shifts the NQR lines by hundreds of kilocycles per second, which is tens or hundreds of times the half-width of the line. The evident conclusion that these frequency shifts are manifestations of relationships well known to chemists, is important. For example (see Table 1), the NQR frequencies in chloro-, bromo-, and iodosubstituted nitrobenzenes are always higher than in the corresponding halogen-substituted anilines, etc. The high precision to which weak manifestations of such polar influences of various substituent groups can be measured by means of NQR is significant. Similar empirical relationships were discovered by Livingston [54] for chloroalkanes. The NQR frequencies for a number of compounds studied by him are shown in Fig. 9.

Livingston attributes the linear increase of frequency in substituted methanes in the CCl_4 direction to the fact that replacement of hydrogen by the more electronegative chlorine makes the C−Cl bond less ionic in character. Further, an examination of the frequency shifts in Fig. 9 leads to the conclusions that replacement of hydrogen by the CH_2Cl group has little effect on NQR frequency, that the CCl_3 group is close to the Cl atom in electronegativity, that the CH_2Cl group has lower electronegativity, that the CH_2Cl group is close to hydrogen in electronegativity, and that CH_3 is more electropositive than hydrogen. It is easily seen in Fig. 9 that with a fixed number of Cl atoms in the molecule, lengthening of the aliphatic chain influences the NQR frequency to an extent which diminishes with the distance between the successive added CH_3 group and the chlorine atom. This leads to the relationship, shown schematically in Fig. 10, between NQR frequency and chain length (number of C atoms).

The interesting experimental fact that ortho-neighbors exert a characteristic influence on NQR of Cl^{35} in chlorobenzenes is examined in [55, 56]. For example, in 1,2,3-trichlorobenzene, the NQR frequencies of the

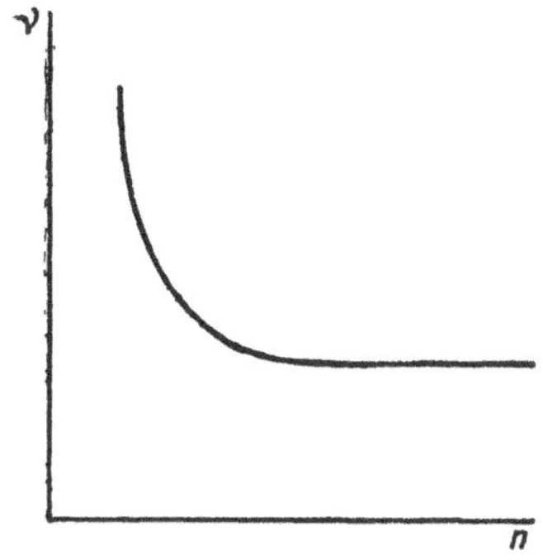

Fig. 10. Variation of the NQR frequency of chlorine with the number n of chlorine atoms in the aliphatic molecule.

chlorine atoms in positions 1 and 3 are lower by 670 kcps than that of the intermediate chlorine atom. The difference between these atoms is evidently that each of the first two has one neighboring Cl atom in the ortho-position, whereas the third has two such Cl atoms. Analogous differences of NQR frequencies are found between Cl atoms without any ortho-neighbors and each with one Cl atom in the ortho position. Duchesne and Monfils [55] attribute these facts to disturbance of the plane character of the chlorobenzene molecule when the Cl atoms are in the ortho position. If the chlorobenzene molecule is plane, then the C–Cl bond has a considerable proportion of π character, and this, as we know, lowers the NQR frequency. According to Duchesne and Monfils, removal of the C–Cl bond from the plane of the benzene ring should diminish this π bonding and hence raise the NQR frequency.

However, Bray et al., showed that, in reality, the double-bond character of the C–Cl bond increases when it is moved through an angle φ out of the ring plane, because the decrease of π-orbit overlapping, which is proportional to $\cos\varphi$, is more than compensated by increased overlapping of the σ orbits, which is proportional to $\sin\varphi$. Then the increase of the NQR frequency on entry of a neighboring Cl atom into the ortho-position should be attributed to a decrease in the ionic character of the C–C σ bond, such as was observed by Livingston for chloroalkanes; this decrease of ionic character should predominate over the frequency decrease caused by increase of double-bond character.

Independent confirmation that the treatment of Duchesne and Monfils is erroneous is provided by the results of an x-ray structural investigation of hexachlorobenzene conducted by Strel'tsova and Struchkov [57]. The molecule of this compound in the crystal was found to be plane, whereas Duchesne and Monfils were forced to ascribe a value of 25° to the angle φ for hexachlorobenzene in order to explain the observed frequency shifts.

This example confirms once again that it is necessary to compare NQR data with results obtained by other methods. However, the question of the form of the hexachlorobenzene molecule and the percentage of double-bond character in the C–Cl bonds can also be solved by the NQR method by investigations of a single crystal in a weak constant magnetic field (see Section 3).

In addition to the influence of ortho-neighbors, Bray et al. [56] showed that the NQR frequency (with a given number of ortho-neighbors) increases linearly with the number of Cl atoms attached to the benzene ring. This result is explained exclusively by the decrease in the ionic character of the C–Cl bond when an additional Cl atom is attached to the ring.

Thus, despite the absence of a precise theory, work on the chemical applications of NQR has yielded much experimental data in a few years, and has revealed a number of interesting empirical relationships. Much work remains to be done in this field. To aid future searches and correlations, the authors of this review have assembled data on all measurements of NQR frequencies in various compounds known to them as of the summer of 1960. Omissions in the appended table of NQR spectra are possible, but they are probably few.* In general, we give the frequencies at 77°K, so as not to make the table too large. If data for some other temperature are given, this means either that no values at 77°K are available, or that some interesting fact associated with the effect of temperature on quadrupole frequencies is present. In several cases we have used this table for successful prediction of unknown NQR frequencies for chlorine and bromine compounds, by comparison with frequencies of analogous substances containing other halogens. In addition, the relationship illustrated in Fig. 10 and a number of other facts have been established by means of this table. Some of these facts are discussed below.

5. NQR Spectra and the Structure of Solids

Even the earliest work on quadrupole resonance revealed the high sensitivity of the method to crystal effects. These effects are: a) splitting of NQR lines owing to nonequivalence of the positions of the molecule in

* The frequencies of more than 100 compounds are given in a fundamental paper by Hooper and Bray [J. Chem. Phys. 33, 334 (1960)].

the lattice; b) line broadening due to crystal-lattice imperfections; c) influence of thermal motion in the crystal on the frequency and form of the line. In the table of NQR spectra, the reader will find numerous examples showing that crystal shifts of NQR frequencies are usually much greater than the line width, and their measurements make no special demands on the resolving power of the instrument.

In chemical investigations such crystal splitting is a hindrance, as it makes interpretation of the results difficult and, in many cases, gives rise to doubt whether the observed effect is of chemical origin. There is an empirical upper limit of crystal effects for every range of quadrupole frequencies. For example, Dewar and Lucken [45] consider that, in the case of NQR of chlorine, frequency shifts up to 500 kcps may be of crystal origin, while all shifts greater than 500 kcps can be, with confidence, ascribed to various chemical factors. Unfortunately, first, it is difficult to find any serious justification for this particular restriction of the range of crystal effects; and, second, in a fairly large number of cases effects of undoubted chemical origin are smaller, such as the chemical shifts in many chlorobenzenes. Estimates with the aid of Eq. (7) (see below) indicate that, apparently, in some cases, we may find crystal splitting of the NQR lines of the order of 10% of the frequency, i.e., 2-3 Mcps in the chlorine band and 10-20 Mcps in the bromine and iodine bands. It may be that it is this that accounts for the splitting observed for $O_2IC_6H_5$ ($\Delta\nu$ = 18.4 Mcps), 2,6-dibromo-4-nitrophenol ($\Delta\nu$ = 13.7 Mcps), $O_2(CHCl)_4$ ($\Delta\nu$ = 2.1 Mcps), and certain other compounds (see Table 1).

It is therefore clear that elucidation of the main relationships between crystal structure and NQR signals is of fundamental importance for putting the chemical applications of quadrupole spectra on a firm basis. It is also quite evident that quadrupole resonance may prove a powerful tool for investigating a number of problems in the physics of solids as such. Here the influence of thermal vibrations and defects in real crystals on NQR is especially important.

It was already stated in the first part of this review that a theory of crystal effects in NQR can be formulated only after the laws of intermolecular interaction have become known. Much work has recently been done in this field, showing that there are real prospects of investigating and understanding certain relationships between intermolecular interaction and quadrupole resonance in molecular crystals.

It may be assumed a priori that the fundamental principles of organic crystal chemistry [58], established on the basis of very extensive experimental data, would prove useful in studies of crystal effects in NQR spectra of organic compounds. The facts support this assumption. On the other hand, it is quite reasonable to hope that the extreme sensitivity of NQR to perfection of the crystal lattice would make it possible to define more closely the concepts of organic crystal chemistry on molecular interaction in crystals.

Influence of Crystal Imperfections on NQR. Many studies have been concerned with the effects of various distortions of the crystal lattice on quadrupole resonance signals. Two methods are used in such investigations: either the crystal is subjected to the action of γ-, β-, or neutron radiation [59-61], or the lattice is deformed by formation of substitutional solid solutions [61-70]. In either case, the amplitude of the NQR signal is lowered.

The rate at which the signal amplitude falls with increase of the absorbed radiation dose is a measure of the radiation stability of the specimen. According to Duchesne [59], the measure of the radiation stability of a crystal to the action of γ-radiation is the dose which reduces the intensity of the NQR signal by 40%. He also showed that the signal amplitude decreases roughly exponentially with increase of the dose. However, this relationship breaks down at high doses; the decrease of the NQR signal slows down and eventually almost ceases with increase of the dose. This observation has been checked and confirmed [60, 61]. Kitaigorodskii and Fedin [60] offered the following explanation for this fact: irreversible chemical changes take place in defective regions of the crystal, and the molecular fragments are packed more loosely during the filling of voids than the molecules in the lattice of a nonirradiated crystal. After all the voids in the crystal have been filled with radiolysis products, the average mobility of the molecular fragments over the crystal drops sharply and further increase of the dose has little effect on the amplitude of the NQR signal. Simple calculations based on this model can be used for estimating block dimensions and the volume percentage of voids in the crystal from experimental data. The concepts used in [60] inevitably led to the conclusion that a carefully grown single crystal should have much higher radiation resistance than an ordinary crystal, and still more than a polycrystalline specimen. This conclusion was experimentally confirmed quite independently by Adler [71], who used a different method. Thus, the theory of close packing of organic molecules in crystals [58] proved fruitful in the solution of certain problems of NQR spectroscopy.

(text continued on p. 113)

TABLE 1

Substance	NQR frequencies, Mcps	T, °K	Literature sources and other data (eqQ in Mcps)
1	2	3	4
	I. Cu^{63-65}		
	Cu – C bond		
K [Cu (CN)$_2$]	ν_1=32.661 ν_2=30.221	298	[83], ν_1 and ν_2 refer to the Cu^{63} and Cu^{65} isotopes, respectively
	Cu – O bond		
$Cu_2^{63}O$	26.697	298	[83]
	II. Hg^{201}		
	Hg – Cl bond		
$Hg^{201}Cl_2$	361.966	83	[84]
	III. B^{10-11}		
B^{11}	2.695	298	[85] $eqQ_{B^{11}} = 5.39$
	B – C bond		
$B^{11}H_3CO$	0.775	298	[85] $eqQ = 1.55$
I.B $(CH_3)_3$	ν_1=1.5225 ν_2=2.5378 ν_3=2.4367	83 83 83	[86]. ν_1 and ν_2 in both I and II refer to B^{10}, ν_3 refers to B^{11}; I and II are two crystalline modifications, with a phase transition, lasting several hours, at 112°K
II.B $(CH_3)_3$	ν_1=1.5536 ν_2=2.5890 ν_3=2.4845	83 83 83	
B $(C_2H_5)_3$	ν_1 { =1.5554 =1.5740 ν_2=2.6068 ν_3=2.5016	83 83 83	[86]. ν_1 (doublet) and ν_2 refer to B^{10}; ν_3 refers to B^{11}
	B – O bond		
$Na_2B_4^{11}O_7 \cdot 4H_2O$	ν_1=1.286 ν_2=1.287	296,89	[78]. $\eta_1 = 16,3\%$; $\eta_2 = 11,7\%$
	IV. Al^{27}		
	Al – Br bond		
$AlBr_3$	ν_1=3.0319 ν_2=3.8326	77 77	[43], ν_1 and ν_2 correspond to the transitions $\pm 1/2 \rightleftharpoons \pm 3/2$ and $\pm 3/2 \rightleftharpoons \pm 5/2$, respectively. η = 72.85%, eqQ = 13.858
	V. Ga^{69-71}		
	Ga – Cl bond		
$GaCl_3$	ν_1=29.060 ν_2=18.315	304,8	[88]. ν_1 and ν_2 refer to Ga^{69} and Ga^{71}, respectively
	Ga – Br bond		
$Ga^{69}Br_3$	26,494	77	[89]
	Ga – I bond		
$Ga^{69}I_3$	21,441	298	[89]
	VI. In^{115}		
	In – I bond		
In I_3	ν_1=37.506 ν_2=26.985 ν_3=36.265 ν_4=51.238 ν_5=63.2	298 298 298 298 298	[89]. η = 69.0%; eqQ = 320.41; ν_1, ν_2, ν_3, and ν_4 correspond to the transitions $\pm 1/2 \rightleftharpoons \pm 3/2$, $\pm 3/2 \rightleftharpoons \pm 5/2$, $\pm 5/2 \rightleftharpoons \pm 7/2$; and $\pm 7/2 \rightleftharpoons 9/2$; ν_5 refers to the "forbidden" $\pm 3/2 \rightleftharpoons \pm 7/2$ transition. Also, see [150].
	VII. N^{14}		
	N – H bond		
NH_3	2.6779	77	[90] $eqQ = 3.5705$

TABLE 1 (continued)

1	2	3	4
ND_3	2.4231	77	[90] $eqQ = 3.2308$
	N − C bond		
$(CH_3)_3N$	3.8954	77	[90] $eqQ = 5.1939$
$(CH_2)_6N_4$	3.4076	77	[91] $\eta = 0$
$CO(NH_2)_2$	$\nu_1 = 2.347$ $\nu_2 = 2.9137$	77	[92] $\eta = 32.3\%$; eqQ = 3.507; ν_1 and ν_2 refer to splitting of levels owing to the high asymmetry. parameter (the spin of N^{14} is 1)
p-$(NH_2)_2C_6H_4$	$\nu_1' = 2.674$ $\nu_1'' = 2.690$ $\nu_1''' = 2.694$ $\nu_2' = 3.1902$ $\nu_2'' = 3.2109$ $\nu_2''' = 3.2129$	77 77	[99]. ν_1 and ν_2 correspond to splitting of levels owing to the high asymmetry parameter; ν', ν'', ν''' appear as the result of the crystallographically non-equivalent positions of the nitrogen atoms; eqQ = 3.910 and η = 26.4% (average values for crystallographic positions)
p-$ClC_6H_4NH_2$	2.837 3.382	77 77	[92]. $eqQ = 4.117$; $\eta = 24.3\%$
p-$BrC_6H_4NH_2$	2.860 3.3417	77 77	]92]. $eqQ = 4.135$; $\eta = 23.1\%$
HCN	3.0223 3.0052	77 77	[93]. $eqQ = 4.0183$; $\eta = 0.85\%$
ClCN	2.400 2.428	77 77	[94]
BrCN	2.5109 2.5203	77 77	[91,95]
ICN	2.5424	299.8	[20,91]
CH_3CN	2.8078 2.7992	77 77	[52]. $eqQ = 3.7380$ [93]. $\eta = 0.46\%$
$CH_2(CN)_2$	$\nu_1 = 3.0154$ $\nu_2 = 2.8670$	77 77	[52]. $eqQ = 3.9216$; $\eta = 7.57\%$
$(CNCl)_3$	$\nu_1 = 3.0445$ $\nu_2 = 3.0799$	77 77	[96] $eqQ = 4.083$; $\eta = 1.7\%$
	VIII. As^{75} As − C bond		
$As(C_6H_5)_3$	$\nu_1 = 98.900$ $\nu_2 = 98.500$	77 77	[97]. ν_1 and ν_2 are due to the crystallographically nonequivalent positions of the arsenic atoms
	As − O bond		
As_2O_3	116.781	77	[98,99]
	As − Cl bond		
$AsCl_3$	78.950	77	[97,100]
	As − Br bond		
$AsBr_3$	63.569	83	[97,101]
	As − I bond		
AsI_3	29.338	77	[97,100]
$AsI_3 \cdot 3S_8$	49.501	77	[100]
	IX. Nb^{95} Nb − O bond		
$KNbO_3$	$\nu_1 = \ldots$* $\nu_2 = 1.335$ $\nu_3 = 2.004$ $\nu_4 = 2.674$	77	[102]. ν_1, ν_2, ν_3, and ν_4 correspond to the transitions: $\pm 1/2 \to \pm 3/2$, $\pm 3/2 \to \pm 5/2$, $\pm 5/2 \to 7/2$, and $\pm 7/2 \to \pm 9/2$;
$KNbO_3$	$\nu_1 = 3.030$ $\nu_2 = 2.085$ $\nu_3 = 2.527$	298	eqQ = 16.0 (77°K) and 23.12 (298°K); η = 0 (77°K) and 80.6% (298°K)

* Here and subsequently, "$\nu = \ldots$" indicates that a line probable on theoretical grounds has not been detected experimentally.

TABLE 1 (continued)

1	2	3	4
	ν_4=3.648		
	X. $Sb^{121-123}$		
	Sb–C bond		
$Sb(C_6H_5)_3$	ν'_1=156.200 ν''_2=152.471 ν'_2= 78.274 ν''_2= 76.912 ν'_3=142.174 ν''_3=138.859 ν'_4= 94.741 ν''_4= 92.267 ν'_5= 47.841 ν''_5= 47.545	77	[103]. Doublets ν_1 & ν_2 correspond to the transitions $\pm3/2 \to \pm5/2$ & $\pm1/2 \to \pm3/2$, associated with Sb^{121}; doublets ν_3, ν_4 and ν_5 correspond to $\pm5/2 \to \pm7/2$, $\pm3/2 \to \pm5/2$ $\pm1/2 \to \pm3/2$, associated with Sb^{123}. Doublets arose owing to crystallographically nonequivalent positions of the antimony atoms. For Sb^{121} lines $(eqQ)' = 520{,}84$; $\eta' = 2{,}6\%$, $(eqQ)'' = 509{,}00$: $\eta'' = 8.0\%$. For Sb^{123} lines $(eqQ)' = 664{,}18$, $\eta' = 1{,}8\%$, $(eqQ)'' = 648{,}56$, $\eta'' = 8{,}3\%$
$Sb(C_3H_7O)_3$	ν_1=167.570 ν_2= 85.804 ν_3=... ν_4=101.111 ν_5=...	77	[103]. ν_1 & ν_2 correspond to the transitions $\pm3/2 \to 5/2$ & $\pm1/2 \to \pm3/2$ for Sb^{121}; $(eqQ)_{121} = 560.66$, $\eta_{121} = 13{,}8\%$ ν_4, to transition $\pm3/2 \to 5/2$ for Sb^{123}.
	Sb – O bond		
Sb_2O_3	ν_1=166.45 ν_2= 83.21 ν_3=151.89 ν_4=101.01 ν_5=50.511	77	[13,104] ν_1 & ν_2 –lines for Sb^{121}; ν_3, ν_4 & ν_5 –lines for Sb^{123}; $(eqQ)_{121} = 554\ 83$; $\eta_{121} \approx 0$; $(eqQ)_{123} = 707.11$; $\eta_{123} \approx 0$
	Sb – S bond		
Sb_2S_3	ν_1=88.66300 ν_2=44.33480 ν_3=... ν_4=53,81785 ν_5=26.91568	303,75	[8]. ν_1 & ν_2 —lines for Sb^{121}; ν_4 & ν_5 —lines for Sb^{123}; $(eqQ)_{121} = 295{,}54679$; $\eta_{121} = 0{,}76\%$; $(eqQ)_{123}$=376,73771; η_{123}=0,77%
	Sb – Cl bond		
$SbCl_3$	ν_1=114.300 ν_2= 59.709 ν_3=104.460 ν_4= 68.648 ν_5= 39.094	83	[8,103 — 105] ν_1 & ν_2 —lines for Sb^{121}; ν_3, ν_4 & ν_5 —lines for Sb^{123}; $(eqQ)_{121}$=383.69; η_{121}=18.9%; $(eqQ)_{123}$=489.34; η_{123}=18.7%
	Sb – Br bond		
$SbBr_3$	ν_1=99.215 ν_2=50.273 ν_3=... ν_4=60.107 ν_5=31.186	77	[8,100] ν_1 & ν_2 —lines for Sb^{121}; ν_4 & ν_5 —lines for Sb^{123}; $(eqQ)_{121} = 343.95$; $\eta_{121} = 10\%$; $(eqQ)_{123}$=422.987; η_{123}=9.5%
	Sb – I bond		
SbI_3	ν_1=... ν_2=... ν_3=... ν_4=... ν_5=25.406	77	[100]. ν_5—line for Sb^{123}, corresponding to the transition $\pm1/2 \to \pm3/2$; $(eqQ)_{123} \approx 169.37$
$SbI_3 \cdot 3S_8$	ν_5=37.461	297	[100]. ν_5 — line for Sb^{123}, corresponding to the transition $\pm 1/2 \to \pm 3/2$; $(eqQ)_{123}$ 249.74
	XI. Bi^{209}		
	Bi – C bond		
$Bi(C_6H_5)_3$	ν_1= 29.785		[106]. ν_1, ν_2, ν_3, and ν_4 correspond

TABLE 1 (continued)

1	2	3	4
	ν_2= 55,214 ν_3= 83,516 ν_4=111,438	83	to the transition $\pm1/2\rightarrow\pm3/2$, $\pm3/2\rightarrow\pm5/2$, $\pm5/2\rightarrow\pm7/2$, $\pm7/2\rightarrow\pm9/2$; eQ=669,06 $\eta\approx9\%$
	Bi–Cl bond		
$BiCl_3$	ν_1=33,726 ν_2=25.905 ν_3=37,900 ν_4=52.704	83	[107]. ν_1, ν_2, ν_3 & ν_4 correspond to the transitions $\pm1/2\rightarrow\pm3/2$, $\pm3/2\rightarrow\pm5/2$, $\pm5/2\rightarrow\pm7/2$ & $\pm7/2\rightarrow\pm9/2$; eqQ=325.5; η=58.3%
	XII. S^{33}		
S_8	22.801 22.866 22.896 22.964	298	[108]. Multiplet character (4 lines) due to nonequivalent positions of sulfur atoms in the crystal
	XIII. Cl^{35-37}		
	Cl–H bond		
HCl^{35}	26,695	20	[54]
	Cl–Hg bond		
$HgCl_2$	ν'_1=22,2303 ν''_1=22,0503 ν'_2=17.5197 ν''_2=17.3789	303	[84]. The doublet ν_1 corresponds to Cl^{35} ; the doublet ν_2 to Cl^{37}
	Cl–B bond		
BCl_3	ν'_1=21.582 ν''_1=21.578 ν'_2=17.008 ν''_2=17,004	77	[40]. The doublet ν_1 corresponds to Cl^{35} ; the doublet ν_2 to Cl^{37}
	Cl–Ga bond		
$GaCl_3$	ν'_1=20,225 ν''_1=19,084 ν'''_1=14.667 ν'_2=15,935 ν''_2=15,040 ν'''_2=11,562	306,2	[88]. ν_1 corresponds to Cl^{35}, ν_2 to Cl^{37}. The multiplet character is due to nonequivalence in the crystal. ν'''_{1-2} is very distinct from ν'_{1-2} and ν''_{1-2} due to dimerization.
	Cl–C bond		
CF_2BrCl	38,342 38.675	77 20	[54]. Cl^{35}
CF_3Cl	38.089 38,790	77 20	[54]. Cl^{35}
$COCl_2$	36.225 36,081	77	[54,63]. Cl^{35} [54]. Cl^{35}
CF_2Cl_2	38.450 39,078	77 20	
CCl_4	40,465 40,521 40.540 40.549 40.576 40,587 40,607 40,639 40.643 40,655 40,696 40.721 40,782 40,797 40.817	77	[54, 109, 110, 111]. Cl^{35}. Multicomponent nature due to nonequivalent positions of chlorine atoms in crystal

TABLE 1 (continued)

1	2	3	4
$CHCl_3$	ν_1=38.254 ν_2=38.308	77	[54, 109, 111]. ν_1 and ν_2 correspond to crystallographically non-equivalent atoms of Cl^{35}
$CHFCl_3$	36.7	20	[112]. Cl^{35}
CHF_2Cl	35.2	20	[112]. Cl^{35}
CH_2FCl	33.799	20	[54]. Cl^{35}
CH_2Cl_2	35.991	77	[54, 109, 11].Signal Cl^{35}
CH_3Cl	ν_1=34.029 ν_2=26.81552	77	ν_1 measured in [54] for Cl^{35}; ν_2 measured in [8] for Cl^{37}
Cl_3CCOCl	ν_1=40.613 ν_2=40.473 ν_3=40.132 ν_4=33.721	77	[113]. ν_1 to ν_3 correspond to chlorine atoms in methyl group; ν_4 to chlorine in COCl group
$ClOCCOCl$	33.621	77	[114]. Cl^{35}
Cl_3CCN	41.7296 41.6660 41.5340	77	[115]. Cl^{35}
Cl_3CCCl_3	40.761 40.714 40.685 40.551	77	[54]. Cl^{35}
	40.885 40.823 40.798 40.652	20	
$ClCF_2COOH$	37.485	77	[116]. Cl^{35}
$CFCl_2COOH$	39.594	77	[116]. Cl^{35}
Cl_2CHCF_3	38.694	77	[116]. Cl^{35}
$Cl_2HCCOCl$	ν_1=39.386 ν_2=39.189 ν_3=38.521 ν_4=38.353 ν_5=32.962 ν_6=32.147	77	[114]. Cl^{35}; $\nu_1 - \nu_4$ correspond to chlorine atoms in methyl group; ν_5 and ν_6 to chlorine atom of COCl. ν_1 and ν_2, ν_3 and ν_4, ν_5 and ν_6 are doublets caused by crystallographic factors
Cl_3CCOOH	40.240 40.165 39.967	77	[113]. Cl^{35}
$Cl_2CHCOOH$	38.807 37.979	77	[113]. Cl^{35}
$CH_2ClCOCl$	ν_1=37.517 ν_2=30.437	77	[114]. Cl^{35}; ν_1 corresponds to chlorine in methyl group; ν_2 —to chlorine of the COCl group
Cl_2CHCN	40.0840 39.9233 39.7537 39.4434	77	[115].Cl^{35}
Cl_2CHCCl_3	39.862 39.836 39.758 38.759 38.711	77	[54]. Cl^{35}
	39.988 39.954 39.923 38.888 38.852	20	
$ClCH_2COONa$	34.794	77	[113]. Cl^{35}
$ClCH_2CN$	38.1251	77	[115]. Cl^{35}
cis-ClCH = CHCl	35.029 34.968	20	[54]. Cl^{35}, see also in [42]
trans -ClCH = CHCl	ν_1 = 35.9862 ν_2 = 28.3646	77	In [8] ν_1 corresponds to Cl^{35}, ν_2 — Cl^{37}. In [54], [56] the NQR

TABLE 1 (continued)

1	2	3	4
	35.584	20	frequency was measured at 20°K; measurements also in [117].
Cl_3CCONH_2	40.007 39.817 39.665 39.599 39,478 38.850	77	[113]. Cl^{35}. At 196°K, only three lines found instead of six; this indicates a phase transition in the 77-196°K range
	39.528 39.229 39,106	196	
$ClCH_2COOH$	36.131 36.429	77	[113]. Cl^{35}
$Cl_2CHCONH_2$	37.750 37,238	77	[118]. Cl^{35}
CH_2-$CHCl$	33.61	20	[56]. Cl^{35}
CH_3CCl_3	38.052 37.829	77	[109]. Cl^{35}
$Cl_3CCH(OH)_2$	39.515 39.429 38.190	77	[113]. Cl^{35}
$ClCH_2CONH_2$	34.882	77	[113]. Cl^{35}
$ClCH_2CH_2Cl$	34.361 34.442	77 20	[54]. Cl^{35}
CH_3CHCl_2	35.55	20	[112]. Cl^{35}
$ClCH_2OCH_2Cl$	32.381 32.587	77	[54]. Cl^{35}
$ClCH_2SCH_2Cl$	34.749 34.526	77	[116]. Cl^{35}
C_2H_5Cl	33.5	20	[112]. Cl^{35}
$ClCH_2OCH_3$	30.181	77	[116]. Cl^{35}
$ClCH_2SCH_3$	33.104	77	[116]. Cl^{35}
CH_2-$CHCH_2Cl$	33.753	86	[116]. Cl^{35}
$ClCH_2COCH_3$	35.075 35.484	77	[113]. Cl^{35}
$ClOCOC_2H_5$	33.858	77	[119]. Cl^{35}
$Cl_3CCH(OH)OCH_3$	38.59 38.74 38.83	90	[120]. Cl^{35}
$CH_3CCl_2CH_3$	34.883 35.046	77 20	[54]. Cl^{35}
$(CH_3)_2NCOCl$	31.8	77	[118]. Cl^{35}
C_3H_7Cl	35.753	86	[116]. Cl^{35}
$CH_3CHClCH_3$	32.1	20	[112]. Cl^{35}
$CH_3CHClOCH_3$	33.453	77	[116]. Cl^{35}
Cl — C — C = O ‖ >O Cl — C — C = O	37.945 38.013	86	[45]. Cl^{35}
$Cl_2CFCOCFCl_2$	39.312 39.104	77	[116]. Cl^{35}
2,4,5.6-Cl_4-1,3-N_2C_4	38.420 38.020 37.068 36.560 36.281	77	[51, 121]. Cl^{35}
$F_3CCCl = CClCF_3$	39.491	77	[116]. Cl^{35}
Cl_4C_4S	37.517	86	[45]. Cl^{35}
Cl — C = C = O ‖ >O H — C — C = O	37.159	86	[45]. Cl^{35}
2,4,6-Cl_3-1,3-N_2C_4H	36.205 36.166	77	[44, 121]. Cl^{35}

TABLE 1 (continued)

1	2	3	4
	35.702		
$ClOC(CH)_2COCl$	30.970	77	[114]. Cl^{35}; crystallographic doublet
	30.380		
$4,6\text{-}Cl_2\text{-}1,3\text{-}N_2C_4H_2$	35.425		
	35.308	77	
	35.132		
	35.102		
$2,5\text{-}Cl_2C_4H_2S$	36.669		[45]. Cl^{35}
	36.696	86	
	36.840		
	36.875		
$ClCH_2COCH_2Cl$	35.943	77	[73]. Cl^{35}
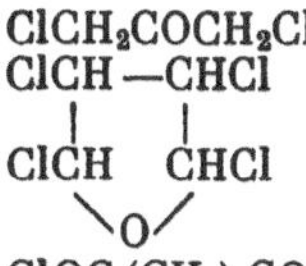	34.824	77	[118]. Cl^{35}
$ClOC(CH_2)_2COCl$	30.217	77	[114]. Cl^{35}
$(C_2H_2Cl_2O)_2$	36.968	77	[118]. Cl^{35}
	34.824		
$Cl_3CCOOC_2H_5$	44.339	77	[119]. Cl^{35}
	40.200		
$2,4\text{-}(NH_2)_2\text{-}6Cl\text{-}1,3\text{-}N_2C_4H$	34.304	77	[121]. Cl^{35}
$CH_2\text{-}CCH_3CH_2Cl$	33.777	77	[116]. Cl^{35}
$ClCH_2COOC_2H_5$	35.962	77	[113]. Cl^{35}
$Cl_3CCH(OH)OC_2H_5$	39.140		[120,122]. Cl^{35}
	38.765	77	
	38.516		
$CH_3CHClOCHClCH_3$	33.392	77	[116]. Cl^{35}
	33.483		
$(CH_3)_3CCl$	31.065	77	[109]. Cl^{35}
	31.195	20	
$1,2\text{-}Cl_2\text{-}3,3,4,4,5,5\text{-}F_6C_5$	38.039	77	[116]. Cl^{35}
$2,4\text{-}Cl_2\text{-}1,3\text{-}N_2C_5H_2$	35.321	77	[51,121]. Cl^{35}
	35.310		
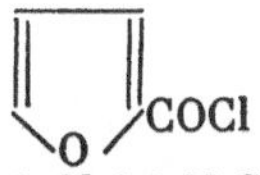	30.726	77	[114]. Cl^{35}
$2\text{-}Cl\text{-}1,3\text{-}N_2C_5H_3$	34.173	86	[44]. Cl^{35}
$2\text{-}ClC_5H_4N$	34.194	77	[123]. Cl^{35}
$3\text{-}ClC_5H_4N$	35.238	77	[51]. Cl^{35}
$4\text{-}ClC_5H_4N$	35.042		[51]. Cl^{35}
	35.031	77	
	34.748		
	34.739		
$3,5\text{-}Cl_2\text{-}C_5H_4N$	35.601	77	[51]. Cl^{35}
$2,4\text{-}Cl_2\text{-}5\text{-}CH_3\text{-}1,3\text{-}N_2C_4H$	35.338		[121]. Cl^{35}
	34.893	77	
$2,4\text{-}Cl_2\text{-}6\text{-}CH_3\text{-}1,3\text{-}N_2C_4H$	35.256	7	[51,121]. Cl^{35}
$4,6\text{-}Cl_2\text{-}2\text{-}CH_3\text{-}1,3\text{-}N_2C_4H$	35.156	8	[44]. Cl^{35}
$2\text{-}Cl\text{-}C_5H_4N \cdot HCl$	37.559	7	[123]. Cl^{35}
$2\text{-}NH_2\text{-}5\text{-}ClC_5H_4N$	35.630	77	[51]. Cl^{35}
$(C_2H_5)_2NCOCl$	31.877	77	[118]. Cl^{35}
$4\text{-}NH_2\text{-}6\text{-}Cl\text{-}2\text{-}CH_3S\text{-}1,3\text{-}N_2C_4H$		77	[51, 121]. Cl^{75}
$6\text{-}Cl\text{-}2,4\text{-}(CH_3O)_2\text{-}1,3\text{-}N_2C_4H$	34,993	77	[51, 121]. Cl^{35}
	34,618		
$6\text{-}NH_2\text{-}2,4\text{-}Cl_2\text{-}C_2H_5\text{-}1,3\text{-}N_2C_4$	34,879	86	[44]. Cl^{35}
	35.044		
$ClOC(CH_2)_4COCl$	28.978	77	[114]. Cl^{35}
$(ClCH_2CHO)_3$	35.363	77	[113]. Cl^{35}
	35.601		
C_6Cl_6	38,403		[55, 124]. Cl^{35}
	38,452	77	
	38,381		

TABLE 1 (continued)

1	2	3	4
$(CNCl)_3$	ν_1=36.7708	77	[36, 44, 51, 110], ν_1, ν_2, and ν_3 refer to Cl^{35}; ν_4 and ν_5 to Cl^{37}.
	ν_2=36.2997		
	ν_3=36.1720		Asymmetry parameters determined
	ν_4=28.6078	294	in [37] at 299°K: η_2 =
	ν_5=28.6208		=26%, η_2 = 23%
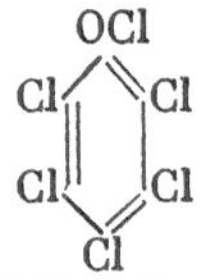	40.100		[122]. Cl^{35}
	39.997	77	
	38.157		
	37.865		
	37.553		
	37.055		
2,3,5,6-$Cl_4C_6O_2$	37.585	77	[51]. Cl^{35}
	37.515		
	37.470		
	37.442		
C_6HCl_5	38.153		[56]. Cl^{35}
	37.959		
	37.720	77	
	37.504		
	37.466		
C_6Cl_5OH	38.5794		[125,126]. Cl^{35}
	38.1915	77	
	37.3750		
	37.1292		
1,2,3,4-$Cl_4C_6H_2$	37.557		[56]. Cl^{35}
	37.455	77	
	37.013		
	37.013		
1,2,3,5-$Cl_4C_6H_2$	37.657		[56]. Cl^{35}
	37.483		
	37.289		
	37.020	77	
	36.872		
	36.816		
	36.386		
	36.185		
1,2,4,5-$Cl_4C_6H_2$	36.898		[127,128]. Cl^{35}
	36.843		
	36.738	77	
	36.702		
2,5-$Cl_2C_6H_2O_2$	36.321	77	[51]. Cl^{35}
2,6-$Cl_2C_6H_2O_2$	36.361		[51]. Cl^{35}
	36.266	77	
2,5-Cl_2-3,6-$(OH)_2$-C_6O_2	37.148	77	[51]. Cl^{35}
2,4,6-$(NO_2)_3$-C_6H_2Cl	39.367	77	[53]. Cl^{35}
1,2,3-$Cl_3C_6H_3$	37.031		[56,129]. Cl^{35}
	36.973		
	36.523	77	
	36.268		
	36.238		
	36.214		
1,2,4-$Cl_3C_6H_3$	36.623		[56,130]. Cl^{35}
	36.400		
	36.380	77	
	36.173		
	35.617		
	35.189		
1,3,5-$Cl_3C_6H_3$	ν_1=36.115		[56]. Cl^{35}. In [37] effect of temperature on NQR frequencies was investigated; found η_1 = 12%, η_2 = 11%
	ν_2=35.894	77	& η_3 = 9% at 299°K
	ν_3=35.545		
2,4,6-$Cl_3C_6H_2OH$	36.770		[53]. Cl^{35}
	35.400	77	
	35.262		

TABLE 1 (continued)

1	2	3	4
2,5-Cl_2-$C_6H_3SO_2Cl$	37,3200 36,3913 34.4484	77	[126]. Cl^{35}
2,4-$Cl_2C_6H_3NO_2$	ν_1=37.874 ν_2=35,921	77	[125, 131]. Cl^{35}. Presumed that ν_1 corresponds to chlorine in the o-position to the NO_2 group, and ν_2 to the m-position
3,4-$Cl_2C_6H_3NO_2$	ν_1=37,055 ν_2=36.488	77	[125]. Cl^{35}. Presumed that ν_1 corresponds to chlorine in the p-position to the NO_2 group, and ν_2 to the m-position
2,4-$(NO_2)_2C_6H_3Cl$	37.796	77	[120,122,132]. Cl^{35}
2,5-$Cl_2C_6H_3SO_2Na$	36,5040 35,2117	77	[126]. Cl^{35}
o-$Cl_2C_6H_4$	35.824 35.580	77	[46]. Cl^{35}
m-$Cl_2C_6H_4$	35.030 35.030 34.875 34.809	77	[46]. Cl^{35}
p-$Cl_2C_6H_4$	ν_1=34,77540 ν_2=27,40811	77	In [8] ν_1 corresponds to Cl^{35} and ν_2 to Cl^{37}. NQR frequencies of p-$Cl_2C_6H_4$ also given in [46, 131, 133]. Asymmetry parameter η = 8% at 299° determined in [37]; effect of temperature on relaxation time T_1 found in [73] (at 77°K T_1 = 560 ±50 sec)
2,6-Cl_2-4-NO_2-$C_6H_2NH_2$	35.920	77	Cl^{35}. Measured by Fedin [127].
2,4,6-$Cl_3C_6H_2NH_2$	35,885 35.780 35,591 35.177 34,976 34.925	77	
o-$ClC_6H_4NO_2$	37.260	77	[14]. Cl^{35}
m-$ClC_6H_4NO_2$	35.457	77	[46]. Cl^{35}
2,6-$(NO_2)_2$-4-Cl-$C_6H_2NH_2$	37.280	77	[53]. Cl^{35}
p-$ClC_6H_4SO_2Cl$	35,967	77	[120]. Cl^{35}. NQR line refers to chlorine in benzene ring.
p-$ClC_6H_4SO_2Na$	35,1366	77	[125.] Cl^{35}
C_6H_5Cl	ν_1=34.6216 ν_2=27.2872	77	[46,54,111]. ν_1 corresponds to Cl^{35}, ν_2 to Cl^{37}
m-ClC_6H_4OH	34.825 34.766	77	[53]. Cl^{35}
p-$Cl_6C_6H_4OH$	34.945 34.700	77	[46]. Cl^{35}
2,4-$Cl_2C_6H_3NH_2$	34.854 34.734	77	[53]. Cl^{35}
3,4-$Cl_2C_6H_3NH_2$	35.872 35.673	77	[53]. Cl^{35}
2,5-$Cl_2C_6H_3NH_2$	34.530 34.413	77	[120,131]. Cl^{35}
2,4,6-$Cl_3C_6H_2NHNH_2$	36.235 35,810 35.569	77	[127]. Cl^{35}
m-$ClC_6H_4NH_2$	34.468 34.388	77	[53]. Cl^{35}
p-$ClC_6H_4NH_2$	34.146	77	[46]. Cl^{35}
p-$ClC_6H_4NH_2$·HCl	35.448	86	[58]. Cl^{35}. NQR of chlorine in HCl not detected
p-$ClC_6H_4NH_2$·HBr	34.752	86	[45]. Cl^{35}

TABLE 1 (continued)

1	2	3	4
2,5-$Cl_2C_6H_3SO_2Cl \cdot 2H_2O$	36.4500 36,1600 35.6222	77	[126]. Cl^{35}
2,6-$Cl_2C_6H_8O$	35,501	77	[134]. Cl^{35}
6,6-$Cl_2C_6H_8O$	ν_1=36.475 ν_2=36.703 ν_3=28,769 ν_4=28.963	77	[134]. ν_1 & ν_2 refer to Cl^{35}; ν_3 & ν_4 to Cl^{37}
2,4-$Cl_2C_6H_3COCl$	ν_1=37.145 ν_2=37.038 ν_3=35.923 ν_4=35.834 ν_5=31.086 ν_6=31.039	77	[114]. Cl^{35}. ν_5 & ν_6 refer to the COCl group
2,4-$Cl_2C_6H_3COOH$	37,432 35,528	77	[53]. Cl^{35}
3,4-$Cl_2C_6H_3COOH$	37,298 36.576	77	[53]. Cl^{35}
o-ClC_6H_4NCO	34,415	196	[46]. Cl^{35}
m-ClC_6H_4NCO	34.653	196	[46]. Cl^{35}
p-$FC_6H_4CCl_3$	39.056 38,772 36.968	77	[118]. Cl^{35}
4-Cl-3-NO_2-C_6H_3COOH	37,843	77	[53]. Cl^{35}
o-$ClC_6H_4CF_3$	35.633	196	[46]. Cl^{35}
m-$ClC_6H_4CF_3$	35.073 34.632	77 196	[46]. Cl^{35}
$C_6H_5CCl_3$	38,898 38,824 38,786 38.713 38.702 38.288	77	[114]. Cl^{35}; crystallographic splitting (two molecules in a common position)
C_6H_5COCl	29;03	77	[118]. Cl^{35}
p-ClC_6H_4CHO	34.623 34.607	77	[46]. Cl^{35}
o-ClC_6H_4COOH	36,3049	77	[126]. Cl^{35}
m-ClC_6H_4COOH	35,227	77	[46]. Cl^{35}
p-ClC_8H_4COOH	34.673	77	[46]. Cl^{35}
5-Cl-2-OH-C_6H_3CHO	34.968	77	[53]. Cl^{35}
p-$ClC_6H_4CH_2Cl$	ν_1=34.303 ν_2=32.804	196	[46]. Cl^{35}; ν_1 refers to chlorine atom in benzene ring; ν_2 to chlorine atom in methyl group
2,4-$Cl_2C_6H_3OCH_3$	36,256 35.734	77	[53]. Cl^{35}
4,6-Cl_2-3-OH-$C_6H_2CH_3$	35.450 34.749	77	[53]. Cl^{35}
2-Cl-6-NO_2-$C_6H_3CH_3$	35.219	77	[53]. Cl^{35}
p-$NO_2C_6H_4CH_2Cl$	34.311	77	[120]. Cl^{35}
4-Cl-3-CH_3-2-NO_2-C_6H_2OH	35.660	77	[53]. Cl^{35}
$C_6H_5CH_2Cl$	33.630	77	[46, 54]. Cl^{35}
p-$ClC_6H_4OCH_3$	34.753	77	[46]. Cl^{35}
4-Cl-3-CH_3OH-C_6H_3-1	34.887 34.535	77	[53]. Cl^{35}
5-Cl-2-CH_3O-$C_6H_3NH_2$	34,220	77	[53]. Cl^{35}
p-$ClC_6H_4COCH_2Br$	34.8224	77	[126]. Cl^{35}
p-$ClC_6H_4CH_2CN$	34;634 34,593	77	[53]. Cl^{35}
p-$ClC_6H_4COCH_3$	34.618	77	[46]. Cl^{35}
p-$ClC_6H_4NHCOCH_3$	34.792	77	[53]. Cl^{35}
2,4-$Cl_2C_6H_3OC_2H_5$	36.108 35,406	77	[53]. Cl^{35}
p-$ClC_6H_4OC_2H_5$	34.381	77	[46]. Cl^{35}
4-Cl-3,5-$(CH_3)_2$-C_6H_2OH	34.415 34,348	77	[126,129]. Cl^{35}

TABLE 1 (continued)

1	2	3	4
4,7-$Cl_2C_9H_5N$	35.591 35.179 34.764	77	[123]. Cl^{35}
6-ClC_9H_6N	34.628	77	[44,123]. Cl^{35}
7-ClC_9H_6N	34.681	86	[44]. Cl^{35}
2-ClC_9H_6N	33.271	77	[44,123]. Cl^{35}
2,4-Cl_2-3-CH_3-C_9H_4N	35.627 34.268	86	[44]. Cl^{35}
2-Cl-C_9H_6N.HCl	37.145	77	[123]. Cl^{35}
p-$ClC_6H_4CH{=}CHCOOH$	34.227	196	[46]. Cl^{35}
$ClOC(CH_2)_8COCl$	29.118	77	[114]. Cl^{35}
2,3-Cl_2-1,4-O_2-$C_{10}H_4$	37.114	77	[51]. Cl^{35}
1,2,3,4-$Cl_4C_{10}H_8$	36.783 36.100 35.262	77	[120]. Cl^{35}
5-C_6H_5-4,6-Cl_2-2-CH_3-1,3-N_2C_5	35.248	86	[44]. Cl^{35}
4,6-Cl_2-2-CH_3-5-C_6H_5-1,3-N_2C_4	35.248	86	[44]. Cl^{35}
$Cl_3CCH(C_6H_4Cl)_2$	ν_1=39.039 ν_2=38.814 ν_3=38.487 ν_4=34.977 ν_5=34.868	77	[114]. Cl^{35}. $\nu_1 - \nu_3$ due to three chlorine atoms in the trichloromethyl group, and ν_4 and ν_5 to chlorine atoms in the benzene rings
	Cl–Si bond		
$HSiCl_3$	19.30	4	[54]. Cl^{35}
$SiCl_4$	20.464 20.415 20.408 20.273	77	[54]. Cl^{35}
$C_2H_5SiCl_3$	18.756 18.842 18.865	300	[132]. Cl^{35}
	Cl–Ti bond		
$TiCl_2$	4.17	297	[135]. Cl^{35}
$TiCl_3$	7.39	297	[135]. Cl^{35}
$TiCl_4$	5.922 5.963 5.988	243	[136]. Cl^{35}
	6.022 6.082 6.150	77	
	Cl–Ge bond		
$GeCl_4$	25.746 25.736 25.714 25.451	77	[54]. Cl^{35}
	Cl–Sn bond		
$SnCl_4$	24.294 24.226 24.140 23.719	77	[54]. Cl^{35}
	Cl–N bond		
$(CH_2CO)_2NCl$	54.100	77	[133]. Cl^{35}
p-OC_6H_4NCl	44.992	77	[133]. Cl^{35}
	Cl–V bond		
VCl_3	9.40	297	[135]. Cl^{35}
	Cl–P bond		
PCl_3	26.202 26.107	77	[54,110]. Cl^{35}
$POCl_3$	ν_1=28.9835 ν_2=28.9378 ν_3=22.8432 ν_4=22.8067	77	[38,54,110] ν_1 & ν_2 refer to Cl^{35}, ν_3 and ν_4 to Cl^{37}

TABLE 1 (continued)

1	2	3	4
$POCl_2F$	29.272 28.715	20	[54]. Cl^{35}
PCl_5	32.630 32.384 32.282	77	[110,118]. Cl^{35}
$(PNCl_2)_3$	27.630 27.704 27.834 27.903	287	[110,137,138]. Cl^{35}
$(PNCl_2)_4$	27.265 28.124 28.182 28.616	287	[137,138]. Cl^{35}
	Cl–As bond		
$AsCl_3$	25.406 25.058 24.960	77	[54, 100]. Cl^{35}
	Cl–Sb bond		
SbCl	ν_1=20.90767 ν_2=19.30468 ν_3=16.47790 ν_4=15.21522	77	[8, 54, 100, 105]. ν_1 and ν_2 refer to Cl^{35}, ν_3 and ν_4 to Cl^{37}
$SbCl_5$	ν_1,ν_2=30.4 ν_3=28.3 ν_4,ν_5,ν_6=27.88	77	[118]. Cl^{35}. ν_1 & ν_2 are a partially allowed doublet; ν_3 is a single line; ν_4, ν_5 and ν_6 is a partially allowed triplet
	Cl–Bi bond		
$BiCl_3$	ν_1=15.849 ν_2=19.531 ν_3=12.498 ν_4=15.387	83	[107]. ν_1 and ν_2 refer to Cl^{35}, ν_3 and ν_4 to Cl^{37}
	Cl–O bond		
$NaClO_2 \cdot xH_2O$	53.33	77	[139]. Cl^{35}
$AgClO_2$	54.08	297	[139]. Cl^{35}. Calculation similar to that in [43] gave asymmetry parameter η = 73%.
$NaClO_3$	ν_1=30.63017 ν_2=24.14228	77	[112]. ν_1 refers to Cl^{35}, ν_2 to Cl^{37}
$KClO_3$	28.954	77	[39, 112, 137] Cl^{35}
$Cu(ClO_3)_2 \cdot 6H_2O$	30.064	77	[114]. Cl^{35}
$AgClO_3$	29.421	77	[112]. Cl^{35}
$Mg(ClO_3)_2 \cdot H_2O$	29.885	77	[114]. Cl^{35}
$Ca(ClO_3)_2$	29.6	290	[140]. Cl^{35}
$Sr(ClO_3)_2$	29.869	77	[114]. Cl^{35}
$Ba(ClO_3)_2 \cdot 6H_2O$	29.922	77	[121. 122]. Cl^{35}
$Zn(ClO_3)_2 \cdot H_2O$	30.026	77	[121, 122]. Cl^{35}
$Ni(ClO_3)_2 \cdot 6H_2O$	30.572	77	[112]. Cl^{35}
	Cl–S bond		
$(SNCl)_3$	29.8	285	[110]. Cl^{35}
$SOCl_2$	ν_1=32.0908 ν_2=31.8874 ν_3=25.2935 ν_4=25.1331	77	[54, 110, 111]. ν_1 & ν_2 are due to Cl^{35}, ν_3 and ν_4 to Cl^{37}
SO_2Cl_2	37.822 37.613	77	[110, 114]. Cl^{35}
$C_2H_5SO_2Cl$	32.5190	77	[126]. Cl^{35}
$CH_3(CH_2)_3SO_2Cl$	32.7592	77	[126]. Cl^{35}
SCH=CHCH=CSO₂Cl (ring: S bonded to C)	33.1507	77	[126]. Cl^{35}
$C_6H_5SO_2Cl$	32.8920 32.5380 32.4700	77	[118, 126]. Cl^{35}
$p\text{-}ClC_6H_4SO_2Cl$	35.967	77	[120]. Cl^{35}, NQR frequency of Cl^{35} in the benzene ring also given

TABLE 1 (continued)

1	2	3	4
2.5-$Cl_2C_6H_3SO_2Cl$	37.3200		[126] Cl^{35}
	36.3913	77	
	34.4484		
p-$BrC_6H_4SO_2Cl$	32.895	77	[120]. Cl^{35}
m-$NO_2C_6H_4So_2Cl$	32.832	77	[120]. Cl^{35}
	32.337		
p-$NO_2C_6H_4SO_2Cl$	33.438	77	[120]. Cl^{35}
p-$CH_3C_6N_4SO_2Cl$	32.460	77	[118]. Cl^{35}
2,5-$(CH_3)_2C_6H_3SO_2Cl$	32.7579		[126]. Cl^{35}
	32.5269	77	
	32.3178		
	32.2698		
	Cl–Cr bond		
$CrCl_2$	8.52	297	[135]. Cl^{35}
$CrCl_3$	ν_1=12.812	297	[135]. Cl^{35}. Intensity
	ν_2=12.848		ratio $A_1 : A_2 = 2 : 1$
	Cl–Se bond		
$SeCl_4$	36.794		[114]. Cl^{35}
	36.171		
	35.696		
	35.041	77	
	34.685		
	34.534		
	Cl–W bond		
WCl_6	10.172	77	[136]. Cl^{35}
	Cl–F bond		
ClF	70.700	20	[54]. Cl^{35}
ClF_3	ν_1 = 75.1295		[54]. ν_1 refers to Cl^{35}, ν_2
	ν_2 = 59.2147	77	to Cl^{37}
	Cl–Cl bond		
Cl_2	ν_1 = 54.2475		[33, 54], ν_1 refers to Cl^{35},
	ν_2 = 42.7544	77	ν_2 to Cl^{37}
	Cl–I bond		
α-ICl	37.185	293	[141]. Cl^{35}
β-ICl	37.202	77	[114]. Cl^{35}
$ICl_3(I_2Cl_6)$	35.680	77	[114]. Cl^{35}
	33.916		
$KICl_2$	19.6272	298,09	[142]. Cl^{35}
	19.0018	298,11	
	18.7645	298,10	
	18.4077	298,10	
$KICl_2 \cdot H_2O$	18.81	296	[142, 143]. Cl^{35}
	19.2070	77	
$CsICl_2$	19.8611	299	[142]. Cl^{35}
	20.0895	77	
PCl_4ICl_2	18.94	293	[143]. Cl^{35}
NH_4ICl_2	26.14	294	[143]. Cl^{35}
$C_5H_5NHICl_2$	17.62	293	[143]. Cl^{35}
$N(CH_3)_4ICl_2$	19.35	295	[143]. Cl^{35}
$NaICl_4 \cdot 2H_2O$	23.2656		[142]. Cl^{35}
	22.8844	298	
	22.1758		
	20.5916		
	23.8274		
	23.2509	77	
	22.6561		
	20.0100		
$KICl_4$	22.38	296	[142, 143]. Cl^{35}
	22.6039	77	
$KICl_4 \cdot H_2O$	20.14		[143]. Cl^{35}
	24.86	296	
	28.17		
	20.3508		[142]. Cl^{35}
	25.0128	77	
	28.5004		

TABLE 1 (continued)

1	2	3	4
$RbICl_4$	22.3904	298	[142]. Cl^{35}
	22.5726	77	
$CsICl_4$	22.5911	298	[142]. Cl^{35}
	22.1991		
	22.8951	77	
	22.6256		
NH_4ICl_4	22.23	295	[143]. Cl^{35}
$NH_4ICl_4 \cdot H_2O$	19.98		[143]. Cl^{35}
	24.68	296	
	27.96		
$C_5H_5NHICl_4$	22.38	294	[143]. Cl^{35}
	XIV. Br^{79-81}		
	Br—Hg bond		
$HgBr_2$	155.47		[144]. Br^{79}
	156.65	90	
	Br—B bond		
BBr_3	$\nu_1=175.293$	Room	[40]. ν_1 & ν_2 refer to Br^{79}, ν_3 & ν_4 to Br^{81}; $\eta=45\%$; $eqQ=$ $=340$
	$\nu_2=175.264$		
	$\nu_3=146.434$		
	$\nu_4=146.411$		
	Br—Al bond		
$AlBr_3(Al_2Br_6)$	$\nu_1=81.815$		[145]. ν_1, ν_2 & ν_3 refer to Br^{81}, ν_4, ν_5 & ν_6 to Br^{79}
	$\nu_2=95.055$		
	$\nu_3=96.426$	77	
	$\nu_4=97.945$		
	$\nu_5=113.790$		
	$\nu_6=115.450$		
	Br—Ga bond		
$GaBr_3(Ga_2Br_6)$	$\nu_1=100.795$		[89]. Lines ν_1-ν_6 refer to Br^{81}, ν_7-ν_{11} to Br^{79}
	$\nu_2=100.840$		
	$\nu_3=140.660$		
	$\nu_4=140.700$		
	$\nu_5=140.790$		
	$\nu_6=140.845$	77	
	$\nu_7=120.602$		
	$\nu_8=120.635$		
	$\nu_9=168.375$		
	$\nu_{10}=168.430$		
	$\nu_{11}=168.560$		
	....		
	Br—In bond		
$InBr_3$ (In_2Br_6)	$\nu_1=86.912$		[89] ν_1-ν_3 refer to Br^{81}, ν_4—ν_6 to Br^{79}
	$\nu_2=106.974$		
	$\nu_3=107.428$	77	
	$\nu_4=104.112$		
	$\nu_5=128.110$		
	$\nu_6=128.608$		
	Br—C bond		
CF_3Br	252.263	77	[147]. Br^{81}
CF_2Br_2	251.624	77	[147]. Br^{81}
	255.681		
$CFBr_3$	256.592	77	[147]. Br^{81}
	262.371		
CBr_4	Several lines from 265 to 268 Mcps	77	[95, 146]. Br^{81}
$CHBr_3$	250.527		[147, 148]. Br^{81}
	250.666	77	
	251.970		
CH_2Br_2	234.805	77	[146, 147]. Br^{81}
	235.601		
CH_3Br	220.981	77	[147]. Br^{81}
CH_3COBr	181.350		[48]. Br^{81}
	180.140		
	179.536	77	
	178.082		
$BrCH_2COOH$	286.72		[48]. Br^{79}
	284.55	77	

TABLE 1 (continued)

1	2	3	4
$BrCH_2CH_2Br$	216.827		[23, 146, 147]. Br^{81}
	218.977	77	
C_2H_5Br	207.790	77	[146—148]. Br^{81}
C_3H_7Br	210.277	77	[147, 148]. Br^{81}
p-$CH_3(CH_2)_2CH_2Br$	ν_1=249.96		[148]. ν_1 refers to Br^{79},
	ν_2=208.86	90	ν_2 to Br^{81}
3,5-$Br_2C_5H_3N$	278.005	77	[123]. Br^{79}
2-BrC_5H_4N	265.213	77	[123]. Br^{79}
C_6Br_6	255.553		[50]. Br^{81}
	255.060	77	
	254.975		
$C_6O_2Br_4$	298.974		[149]. Br^{79}
	298.400	77	
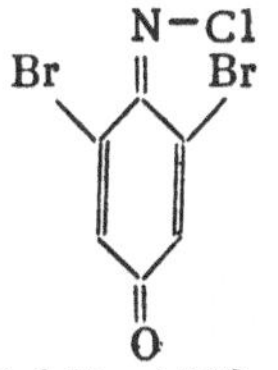	246.736		[133]. Br^{81}
	246.174	77	
2,6-Br_2-4-NO_2-C_6H_2ONa	270.32		[150]. Br^{79}
	270.71	298,5	
1,2,3,5-$Br_4C_6H_2$	236.042		[49, 50] Br^{81}
	242.176		
	242.767	77	
	244.949		
	246.140		
	249.314		
1,2,4,5-$Br_4C_6H_2$	242.75	77	[49]. Br^{81}
	242.00		
2,4-$(NO_2)_2C_6H_3Br$	ν_1=301.270	77	[49] ν_1 refers to Br^{79}, ν_2 to
	ν_2=251.690		Br^{81}
2,6-Br_2-4-NO_2-C_6H_2OH	280.47	296,9	[150]. Br^{79}
	281.16	296,9	
	294.24	297,3	
	294.80	297,3	
2,5-$Br_2C_6H_3NO_2$	307.1		[146]. Br^{79}. Measurements with
	287.2	83	a resonance wavemeter, i.e., of low precision
1,2,4-$Br_3C_6H_3$	240.67		[49]. Br^{81}
	239.47		
	236.37	77	
	232.34		
1,3,5-$Br_3C_6H_3$	234.52		[49, 146]. Br^{81}
	234.11	77	
	233.14		
2,4,6-$Br_3C_6H_2OH$	275.96		[140]. Br^{79}
	277.17	297,3	
	284.40		
o-$BrC_6H_4NO_2$	300.3	83	[146]. Br^{79}. Measurements with a resonance wavemeter; error may reach several Mcps
m-$BrC_6H_4NO_2$	284.4	83	[146]. Br^{79}. Measurements with a resonance wavemeter, i.e., of low precision
p-$BrC_6H_4NO_2$	274.84	299	[150]. Br^{79}
	284.0	83	
o-$Br_2C_6H_4$	236.068		[50]. Br^{81}
	235.969	77	
p-$Br_2C_6H_4$	226.49		[49]. ν_1 refers to Br^{81},
	271.125	77	ν_2 to Br^{79}; [32]. η=5%
2,6-Br_2-4-NO_2-$C_6H_2NH_2$	282.458	77	[149]. Br^{79}

TABLE 1 (continued)

1	2	3	4
2,4,6-$Br_3C_6H_2NH_2$	ν_1=233,56 ν_2=230,75 ν_3=230.10 ν_4=276.00 ν_5=272,68 ν_6=272,29	77 296,8	$\nu_1 - \nu_3$ refer to Br^{81} [49], $\nu_4 - \nu_6$ to Br^{79} [150]
C_6H_5Br	ν_1=224,61 ν_2=268,856	77	[49, 148, 151]. ν_1 refers to Br^{81}, ν_2 to Br^{79}
o-BrC_6H_4OH	275.884	77	[48]. Br^{79}
m-BrC_6H_4OH	272.745 271.420	77	[48]. Br^{79}
p-BrC_6H_4OH	ν_1=269,683 ν_2=268.222 ν_3=225,30 ν_4=224.08	77	[49, 147, 150, 154] ν_1 & ν_2 refer to Br^{79}, ν_3 and ν_4 to Br^{81}
o-$BrC_6H_4NH_2$	220.26	77	[49, 152]. Br^{81}
m-$BrC_6H_4NH_2$	222.61	77	[49. 152]. Br^{81}
p-$BrC_6H_4NH_2$	ν_1=265.525 ν_2=221.86	77	ν_1 refers to Br^{79} [149], ν_2 to Br^{81} [49]. See also [146, 150, 151]
2,6-Br_2-4-NH_2-C_6H_2OH	274.79 277.80	297,7	[150]. Br^{79}
2-Br-3-NO_2-C_6H_3COOH	302.54	296,7	[150]. Br^{79}
C_6H_5COBr	191.868	77	[48]. Br^{81}
o-BrC_6H_4COOH	280.66	298	[20]. Br^{79}
m-BrC_6H_4COOH	269.23	298	[150]. Br^{79}
p-BrC_6H_4COOH	280.08	297,8	[150]. Br^{79}
2-OH-5-Br-C_6H_3COOH	ν_1=271.820 ν_2=227.08	77	[49]. ν_1 refers to Br^{79}, ν_2 to Br^{81}
2,4,6-$Br_3C_6H_2OCH_3$	ν_1=289.194 ν_2=283.888 ν_3=280.632 ν_4=285.800 ν_5=279.980 ν_6=277.255	77 299	$\nu_1 - \nu_3$ to Br^{79} [49, 150], $\nu_4 - \nu_6$ to Br^{81}
o-$BrC_6H_4OCH_3$	233.25	77	[49]. Br^{81}
p-$BrC_6H_4OCH_3$	226.69	77	[49]. Br^{81}
2-CH_3-4-$BrC_6H_3NH_2$	ν_1=263.616 ν_2=220.23	77	[49]. ν_1 refers to Br^{79}, ν_2 to Br^{81}
2-Br-4-$CH_3C_6H_3NH_2$	265.974 263.222	77	[48]. Br^{79}
p-$BrC_6H_4COCH_3$	230.37	77	[49] Br^{81}
m-$BrC_6H_4COOCH_3$	ν_1=271.772 ν_2=227.04	77	[49]. ν_1 refers to Br^{79}, ν_2 to Br^{81}
p-$BrC_6H_4COOCH_3$	ν_1=275.960 ν_2=230.54	77	[49], ν_1 — Br^{79}, ν_2 — Br^{81}
p-$BrC_6H_4NH(COCH_3)$	226.80	77	[49]. Br^{81}
p-$(BrCH_2)_2C_6H_4$	257.697	77	[149]. Br^{79}
p-$(BrCH_2)C_6H_4CH_3$	254.290	77	[49]. Br^{79}
2-Br-4-$CH_3C_6H_3CH_3$	263.332	77	[48]. Br^{79}
p-$BrC_6H_4OC_2H_5$	266.818	77	[48]. Br^{79}
p-$BrC_6H_4N(CH_3)_2$	254.20 255.10 256.03	299	[150]. Br^{79}
3-BrC_9H_6N	275.187	77	[133]. Br^{79}
5,7—Br_2—C_9H_6N—8—OH	282.676 279.960	77	[51]. Br^{79}
1,6-Br_2—2-OH—$C_{10}H_5$	267.87 265.15	299	[150]. Br^{79}
α-$C_{10}H_7Br$	267.895	77	[44]. Br^{79}
p-$BrC_6H_4N(C_2H_5)_2$	263.444 262.906	77	[48]. Br^{79}
4,4'-$BrC_6H_4C_6H_4Br$	271.835 271.110 269.461 267,439	77	[49, 147]. Br^{79}

TABLE 1 (continued)

1	2	3	4
p-$BrC_6H_4(C_6H_5)$	268.70	298,7	[150]. Br^{79}
p-$BrC_6H_4OC_6H_5$	272.500	77	[48]. Br^{79}
2,4,6-$Br_3C_6H_2SO_2C_6H_4$-CH_3-4′	ν_1=297.06		[150]. Br^{79}
	ν_2=288.40	297,3	
	ν_3=281.12		
9,10-$Br_2C_{14}H_8$	ν_1=275.053		[49], [150] ν_1 & ν_2 refer to Br^{79}
	ν_2=272.975	77	
	271.59	297,5	
$C_6H_5CH{=}CBrCOC_6H_5$	276.30	297,8	[150]. Br^{79}
	Br–Si bond		
$SiBr_4$	147.576	77	[95]. Br^{81}
	147.511		
	Br–Ti bond		
$TiBr_4$	47,127	77	[136]. Br^{79}
	46.309		
	Br–Ge bond		
$GeBr_4$	175.949	77	[95]. Br^{81}
	175.584		
	Br–Sn bond		
$SnBr_4$	165.270		[23, 95, 148]. Br^{81}
	165.401	77	
	165.208		
	162.032		
	Br–P bond		
PBr_3	ν_1=220.575		[148, 153]. ν_1 & ν_2 refer to Br^{79}, ν_3 & ν_4 to Br^{81}
	ν_2=218.635	83	
	ν_3=184.257		
	ν_4=182.636		
	Br–As bond		
$AsBr_3$	173.143		[97, 104]. Br^{81}
	172.728	77	
	171.480		
	Br–Sb bond		
$SbBr_3$	ν_1=171.576		[95, 100, 153] ν_1—ν_3 refer to Br^{79}, ν_4—ν_6 to Br^{81}
	ν_2=164.768		
	ν_3=164.492	77	
	ν_4=143.324		
	ν_5=137.630		
	ν_6=137.403		
	Br–O bond		
$LiBrO_3$	184.12	293	[154]. Br^{79}
$NaBrO_3$	ν_1=178.95	293	[95, 154]. ν_1 — Br^{79}, ν_2 — Br^{81}
	ν_2=151.909	77	
$KBrO_3$	ν_1=173.11	293	[95, 154]. ν_1 — Br^{79}
	ν_2=149.361	77	ν_2 — Br^{81}
$CsBrO_3$	ν_1=142.735	296,9	[95]. ν_1 — Br^{81}
	ν_2=145.662	77	ν_2 — Br^{79}
$Cu(BrO_3)_2\cdot 6H_2O$	175.70	285	[154]. Br^{79}
$AgBrO_3$	168.19	293	[154]. Br^{79}
$Mg(BrO_3)_2\cdot H_2O$	176.60	293	[23, 154] Br^{79}
$Ca(BrO_3)_2\cdot H_2O$	176.85	293	[154]. Br^{79}
$Sr(BrO_3)_2\cdot H_2O$	175.25	293	[154]. Br^{79}
$Ba(BrO_3)_2\cdot H_2O$	173.69	293	[154]. Br^{79}
$Zn(BrO_3)_2\cdot 6H_2O$	177.38	285	[154]. Br^{79}
$Cd(BrO_3)_2\cdot H_2O$	174.71	290	[154]. Br^{79}
$HgBrO_3$	168.91	285	[154]. Br^{79}
$Hg(BrO_3)_2$	170.78	285	[154]. Br^{79}
	172.62		
$Pb(BrO_3)_2\cdot H_2O$	174.30	285	[154]. Br^{79}
$Co(BrO_3)_2\cdot H_2O$	176.92	290	[154]. Br^{79}
$Ni(BrO_3)_2\cdot 6H_2O$	177.62	290	[154]. Br^{79}
	Br–Cr bond		
$CrBr_3$	85.73	297	[135]. Br^{81}
	Br–Br bond		
Br_2	ν_1=382.04		[155]. ν_1 — Br^{79}, ν_2 — Br^{81}
	ν_2=319.03	83	η = 20% at 253°K

TABLE 1 (continued)

Substance	NQR frequencies, Mcps transition $\pm\frac{1}{2} \rightleftharpoons \pm\frac{3}{2}$	 transition $\pm\frac{3}{2} \rightleftharpoons \pm\frac{5}{2}$	Temp. °K	Literature sources and other data (eqQ in Mcps)
1	2	3	4	5
	XV. I^{127}			
	I–Hg bond			
K_2HgI_4	148.49 129,30 123.46		293 296	[144]
	I–B bond			
$BI_3(B_2I_6)$	212,6	340.1	196	[22]. $\eta=45.6\%$, $eqQ=1176$. Both lines are doublets with 25 ± 5 kcps splitting
	I–Al bond			
$AlI_3\ (Al_2I_6)$	$\nu_1=112.314$ $\nu_2=131,371$ $\nu_3=131.844$	$\nu_1=215.614$ $\nu_2=263.228$ $\nu_3=263,920$	77	[156]. $\eta_1=18.1\%$, $\eta_2=\eta_3=0$, $(eqQ)_1=723.38$; $(eqQ)_2=876.52$ $(eqQ)_3=879,33$
	I–Ga bond			
$GaI_3\ (Ga_2I_6)$	$\nu_1=135,719$ $\nu_2=176.496$ $\nu_3=177.438$	$\nu_1=253.347$ $\nu_2=352.948$ $\nu_3=354,519$	77	[156] $\eta_1=23.7\%$, $\eta_2=0.9\%$ $\eta_3=2.8\%$; $(eqQ)_1=853.92$, $(eqQ)_2=1176$, $(eqQ)_3=$ $=1186.91$
	I–In bond			
$InI_3\ (In_2I_6)$	$\nu_1=...$ $\nu_2=176.763$ $\nu_3=177,146$ $\nu_4=122.728$ $\nu_5=173.177$ $\nu_6=173,633$	$\nu_1=...$ $\nu_2=353.466$ $\nu_3=354,580$ $\nu_4=229.190$ $\nu_5=346.289$ $\nu_6=347.234$	77 297	[150, 156] $\eta_4=23.7\%$, $\eta_5=1.1\%$, $\eta_6=0$; $(eqQ)_4=772.25$, $(eqQ)_5=1154.33$, $(eqQ)_6=1157.46$
	I–C bond			
CF_3I	310.375		83	[153]
CI_4	319.549		77	[153]
$CHI_3 \cdot 3S_8$	307.10	613.949	77	[153]. $\eta=1.82\%$
CH_2I_2	286.883	568.362	77	[146, 147, 153] $\eta=2.71\%$
CH_3I	265.102	529.670	77	[20, 146, 147, 151, 153] $\eta=2.8\%$
trans-ICH =CHI	277.1	553.82	83	[20] $\eta=2.3\%$
CH_3CHI_2	260,29		90	[78]
C_2H_5I	247,374		77	[146, 147]
ICH_2CH_2COOH	260.9		90	[151]. Measurements with a resonance wavemeter, i.e., of low precision
C_3H_7I	250.962		77	[146]
$(CH_3)_2CHI$	235.921		77	[146]
$o\text{-}C_6H_4I_2$	286,923 286.285 285.531 284.036		 77	[56]

TABLE 1 (continued)

1	2	3	4	5
m-$C_6H_4I_2$	286.3		83	[146]. Measurements with a resonance wavemeter, i.e., with a low precision
p-$C_6H_4I_2$	ν_1=280.147 ν_2=275.89 ν_3=276.31	 ν_2=550.16 ν_3=551.68	77 301	[56, 95, 146]. η_2=4,77% and η_3=3,65%
o-$IC_6H_4NO_2$	ν_1=315.19 ν_2=315.97	ν_1=613.88 ν_2=615.26	299,8	[150]. η_1=14.47% and η_2=14,57%
m-$IC_6H_4NO_2$	ν_1=285,63 ν_2=286.66	ν_1=570.29 ν_2=571.43	300	[150, 151]. η_1=5,05% & η_2=3,65%
p-$IC_6H_4NO_2$	284.36	565.82	300,7	[150]. η=6.30%
C_6H_5I	274.465 274,978		77	[147, 150]
o-IC_6H_4OH	277,60	551.80	298,7	[150.] η=6.91%
p-IC_6H_4OH	278.1		83	[146]. Measurements with a resonance wavemeter, i.e., of low precision
$C_6H_5IO_2$	ν_1=162.23	ν_1=311.70 ν_2=293,30	294 301,4	[150]. η_1=17.89%
o-$IC_6H_4NH_2$	270.0		83	[147]. Measurements with a resonance wavemeter, i.e., of low precision
p-$IC_6H_4NH_2$	ν_1=264.280 ν_2=261.14	 ν_2=521.25	77 300,7	[146, 149, 150]. η_2=3,91%
p-$IC_6H_4NH_2\cdot HCl$	288.28	572,15	296	[150]. η=7,71%
2,5-$I_2C_6H_3CF_3$	287.2		83	[146]. Measurements with a resonance wavemeter, i.e., of low precision
o-IC_6H_4COONa	294,81		296,3	[150]
m-IC_6H_4COONa	ν_1=272,73 ν_2=273.76	ν_1=545,17 ν_2=547.42	298	[150] η_1=2.03% & η_2=1.27%
p-IC_6H_4COONa	277.59 280,82 282.25		295,7	[150]
3.5-I_2-2-OH-C_6H_2-COOH	295.14 284.19		296,6	[150]
o-IC_6H_4COOH	294.58	573.43	300,2	[150] η=14.62%
m-IC_6H_4COOH	275.67	548.56	301	[150]. η=6,27%
p-IC_6H_4COOH	272,89	544.18	300,1	[150]. η=4.76%
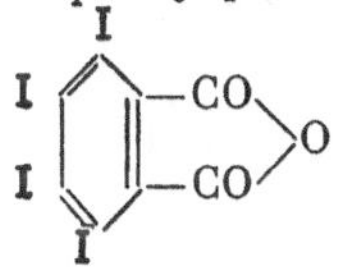	313.63 316,20 324.03 328.53	Three lines about 604, 609, and 616	301,8	[150].
m -$C_6H_4COOCH_3$	276.24	550.16	297	[150]. η=5.71%
p-$IC_6H_4COOCH_3$	284.03	564.21	297	[150].η=7.26%
p-$IC_6H_4NHCOCH_3$	273.74	546,06	301,2	[150]. η=4.49%
p-$IC_6H_4C_6H_5$	275.40		300,4	[150].
	I–Si bond			
SiI_4	ν_1=199.999 ν_2=198.736	ν_1=399.993 ν_2=397.430	77	[153]. η_1=0.33% & η_2=0,91%
	I–Ge bond			
GeI_4	ν_1=225.088 ν_2=222.672	ν_1=450.173 ν_2=445.294	77	[153]. η_1=0,224% & η_2=0,93%

TABLE 1 (continued)

1	2	3	4	5
	I–Sn bond			
SnI_4	$\nu_1=209.133$ $\nu_2=207.683$	$\nu_1=418.257$ $\nu_2=415.320$	77	[153]. $\eta_1=0.4\%$ & $\eta_2=0.92\%$; see also [20, 23, 95, 100]
$SnI_4 \cdot 2S_8$	$\nu_1=210.746$ $\nu_2=207.458$	$\nu_1=420.675$ $\nu_2=414.907$	77	[100]. $\eta_1=3.9\%$, $\eta_2=0.2\%$ $(eqQ)_1=1402.68$ $(eqQ)_2=1383.03$
$SnI_4 \cdot 4S_8$	$\nu_1=204.64$ $\nu_2=205.25$ $\nu_3=210.99$ $\nu_4=219.04$	$\nu_1=409.35$ $\nu_2=410.50$ $\nu_3=421.40$	77	[100]. $\eta_1=\eta_2=0$, $\eta_3=3\%$, $\eta_4=?$, $(eqQ)_1=1364.3$, $(eqQ)_2=1368.3$ $(eqQ)_3=1405.0$, $(eqQ)_4=?$
	I–As bond			
AsI_3	207.023	395.979	77	[100, 101]. $\eta=18.91\%$ & $eqQ=1330.23$
$AsI_3 \cdot 3S_8$	227.543	455.059	77	[100, 153]. $\eta=0.7\%$; $eqQ=1516.86$
	I–Sb bond			
SbI_3	174.356	254.637	77	[95, 97, 100, 103] $\eta=56.5\%$; $eqQ=895.83$
$SbI_3 \cdot 3S_8$	184.151	367.023	77	[100, 153]. $\eta=3\%$; $eqQ=1226.26$
	I–Bi bond			
BiI_3	111.320	201.380	77	[103]. $\eta=29\%$; $eqQ=682$
	I–O bond			
HIO_3	203.12	330.84	77	[112, 150, 160]. $eqQ=1141.0$ $\eta=43.37\%$. A forbidden $\pm 1/2 \rightleftharpoons \pm 5/2$ transition was observed; its frequency was 528, 746 Mcps; (297°K)
KIO_3	$\nu_1=144.37$ $\nu_2=144.91$ $\nu_3=145.38$	$\nu_1=288.82$ $\nu_2=289.85$ $\nu_3=290.71$	299	[150, 157]. $\eta_1\leqslant0.02\%$, $\eta_2\leqslant0.08\%$ and $\eta_3\leqslant0.08\%$
$Ca(IO_3)_2 \cdot 6H_2O$	150.75 150.97	297.470 298.709 307.612	297	[90]
	I–Cl bond			
I_2Cl_6	458.19	909.37	77	[20, 24, 158]. $\eta=7.72\%$; $eqQ=3034.9$
$KICl_2$	469.160 462.060		298	[158].
$KICl_2 \cdot H_2O$	471.051		298	[158].
$RbICl_2$	463.988		298	[158].
$CsICl_2$	464.949		298	[158].
$NaICl_4 \cdot 2H_2O$	461.702		298	[158].

1	2	3	4	5
$KICl_4$	477,506		298	[158].
$KICl_4 \cdot H_2O$	458,863		298	[158].
$RbICl_4$	470,617 458,988		298	[158].
$CsICl_4$	453.123		298	[158].
	I–I bond			
NH_4I_3	366.23	732.13	298	[24]. Doublet line with splitting of several kcps; η = 1.8% eqQ=2440.6
I_2	332.4	644.8	293	[20]. η=15%; eqQ = 2153
	XVI. Co^{59} Co–C bond			
$[C_5H_5CoC_5H_5]$ $[ClO_4]$	ν_1= . . . ν_2= . . . ν_3=24.08 ν_4=24.50 ν_5=36.12 ν_6=36.74		77	[159]. $\eta \approx 0$. Presumed that ν_3 & ν_6 correspond to transition $\pm 3/2 \rightarrow \pm$ $\pm 5/2$, & ν_3, ν_4 to transition $\pm 5/2 \rightarrow \pm 7/2$

Randall, Molton, and Ard [61] showed that polar factors cannot be responsible for the observed decrease of NQR frequency even in irradiated ionic crystals. This shows, once again, that consideration of geometric factors must play an important, if not the decisive role in analysis of the influence of ionizing radiation on quadrupole resonance.

The interesting fact that the rate of phase transition is retarded sharply in p-dichlorobenzene after irradiation by soft γ rays has been reported [72]. Rate of phase transition is calculated on the change in the NQR lines for the appropriate α and β phases. The retardation of phase transition under the influence of radiation indicates that fairly large voids must be present in the lattice for transition to take place; it is difficult to imagine any other mechanism in which filling of the voids in a crystal with radiolysis products can influence rearrangement of the undamaged regions of the lattice.

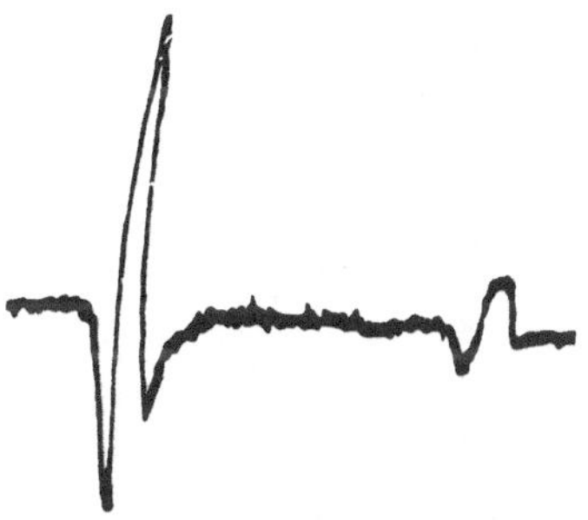

Fig. 11. NQR signals of Cl^{35} from the α and β phases in 0.8 g of p-dichlorobenzene. The distance between the peaks corresponds to 26.4 kcps. The frequency decreases from left to right.

Studies of solid solutions [61-70] occupy an important position among the applications of quadrupole resonance. Monfils [65] was the first to propose that the influence of an impurity molecule in a solid solution on the NQR signal should be characterized by the number N of neighboring molecules of the principal substance which are brought out of resonance by the impurity molecule. At low impurity concentrations it may be assumed that N is a property of the impurity molecule in the given lattice, and is independent of the concentration. It is then easy to show [68] that the signal intensity decreases with increasing impurity concentration c, as follows

$$A = A_0 \, e^{-Nc}, \qquad (6)$$

where A_0 is the amplitude of the signal for the pure substance. Three types of A–c relationships are observed in practice [68] (Fig. 12). Curve 1 is characteristic for unlimited solubility in the given range of concentrations c; Curve 2 indicates limited solubility, and Curve 3 represents the case when the form and dimensions of the impurity molecule prevent formation of a solid solution. It was shown [64,67] that the decrease in the amplitude of the NQR signal is a consequence of the corresponding broadening of the line for the solid solution [see Eq. (3)].

Baer and Dean [70] measured shifts of NQR frequencies in impurity molecules. Their published data are very interesting and merit further study. Fedin and Kitaigorodskii [68] used a very precise method to demonstrate the absence of appreciable shifts of NQR frequencies for molecules of the main substance in solid solutions of a number of compounds in p-dichlorobenzene. They also attempted to use data of organic crystal chemistry for

TABLE 2

List of Substances for Which NQR Signals Could Not be Detected

Substance	Atom	Search band, Mcps	Temp., °K	Literature
$CHBrI_2$	Br^{81-79}, I^{127}	—	83	[146]
$BrCH_2 — CHBr_2$	Br^{81-79}	—	83	[146]
$Br_2CH — CHBr_2$	Br^{81-79}	—	83	[146]
CH_3CH_2COCl	Cl^{35}	27—31	77	[114]
$CH_3(CH_2)_4COCl$	Cl^{35}	24—36	77	[114]
$CH_3(CH_2)_6COCl$	Cl^{35}	25—36	77	[114]
$CH_3(CH_2)_7COCl$	Cl^{35}	24—36	77	[114]
$CH_3(CH_2)_{10}COCl$	Cl^{35}	24—40	77	[114]
$CH_3(CH_2)_{12}COCl$	Cl^{35}	24—36	77	[114]
$CH_3(CH_2)_{14}COCl$	Cl^{35}	24—36	77	[114]
$(CH_3)_2CHCH_2COCl$	Cl^{35}	24—36	77	[114]
$(CH_3)_2CH(CH_2)_2COCl$	Cl^{35}	24—36	77	[114]
$CH_3CH_2CH(C_2H_5)COCl$	Cl^{35}	25—35	77	[114]
$C_6H_5CH_2COCl$	Cl^{35}	24—36	77	[114]
$C_6H_5(CH_2)_2COCl$	Cl^{36}	25—35	77	[114]
$C_6H_5CCH_2COCl$	Cl^{35}	24—36	77	[114]
α-$C_6H_5OCH(CH_3)COCl$	Cl^{35}	24—36	77	[114]
m-$NO_2C_6H_4COCl$	Cl^{35}	27—40	77	[114]
o-$NO_2C_6H_4COCl$	Cl^{35}	27—40	77	[115]
p,5-$(NO_2)_2C_6H_4COCl$	Cl^{35}	27—40	77	[115]
3-ClC_6H_4COCl	Cl^{35}	27—40	77	[115]
3,4-$Cl_2C_6H_3COCl$	Cl^{35}	24—36	77	[115]
m-BrC_6H_4COCl	Cl^{35}	28—40	77	[115]
$CH_3OC_6H_5COCl$	Cl^{35}	24—36	77	]115]
o-$C_2H_5OC_6H_4COCl$	Cl^{35}	24—40	77	[115]
p-$C_2H_5OC_6H_4COCl$	Cl^{35}	27—40	77	[115]
p-$CH_3(CH_2)_4OC_6H_4COCl$	Cl^{35}	27—40	77	[115]
$(C_6H_5)_2NCOCl$	Cl^{35}	27—40	77	[115]
CH_3CH-$CHCOCl$	Cl^{35}	24—36	77	[115]
α-$C_{10}H_7COCl$	Cl^{35}	24—36	77	[115]
2-NH_2-4-Cl-6-CH_3-$C_4H_2N_2$	Cl^{35}	33—39	77	[115]
		31—36	300	
2-NH_2-4,6-Cl_2-5-CH_3-$C_4H_2N_2$	Cl^{35}	33—39	77	
		33—37	300	[134]
2-NH_2-5-Cl-4-HOOC-$C_4H_2N_2$	Cl^{35}	34—39	77	
		33—39	300	[134]
$RbICl_2$	Cl^{35}	16—29	77	[142]
		14—39	298	
m-$Br_2C_6H_4$	Br^{81-79}	—	83	[147]
o-$BrC_6H_4CH_3$	Br^{81-79}	—	83	[147]
m-$BrC_6H_4CH_3$	Br^{81-79}	—	83	]147]
p-$BrC_6H_4CH_3$	Br^{81-79}	—	83	[147]
o-BrC_6H_4I	Br^{81-79}; I^{127}		83	[147]
p-BrC_6H_4I	Br^{81-79}; I^{127}		83	[147]
		200—340	300	[150]
m-BrC_6H_4Cl	Br^{81-79}	—	83	[147]
p-BrC_6H_4Cl	Br^{81-79}	—	83	[147]
2-Br-6-CH_3O-$C_{10}H_6$	Br^{81-79}	200—330	300	[150]
2-Br-6-OH-$C_{10}H_6$	Br^{81-79}	190—320	300	[147]
1-Br-2-OH-$C_{10}H_6$	Br^{81-79}	200—320	300	[150]
1,2-$(CH_3)_2C_6Br_4$	Br^{81-79}	230—300	77	[149]
1,3-$(CH_3)_2C_6Br_4$	Br^{81-79}	230—300	77	[149]
o-$IC_6H_4CH_3$	I^{127}	—	83	[147]
p-$IC_6H_4CH_3$	I^{127}	—	83	[147]
		215—350	300	[150]
p-IC_6H_4Cl	I^{127}	—	83	[147]
		220—370	300	[150]
2,4,6-$I_3C_6H_2COOH$	I^{127}	230—340	300	[150]
2,3,5-$I_3C_6H_2COOH$	I^{127}	240—360	300	[150]
3,5-I_2-4-OH-C_6H_2COOH	I^{127}	240—360	300	[150]
2-NH_2-3.5-I_2-C_6H_2COOH	I^{127}	240—346	300	[150]
o-IC_6H_4COOH	I^{127}	195—360	300	[150]
$(CH_2)_6N_4 \cdot C_2H_5$	I^{127}	200—340	300	[150]
$(C_6H_5)_3N \cdot HI$	I^{127}	155—340	300	[150]

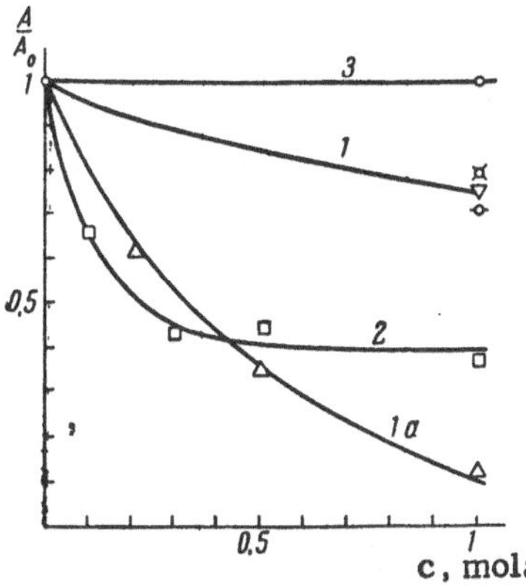

Fig. 12. Variation of the NQR signal amplitude with the impurity concentration in p-dichlorobenzene. Curve 1) unlimited solubility: a) impurity molecule larger than main molecule; b) impurity molecule smaller than main molecule. Curve 2) limited solubility; Curve 3) no solubility.

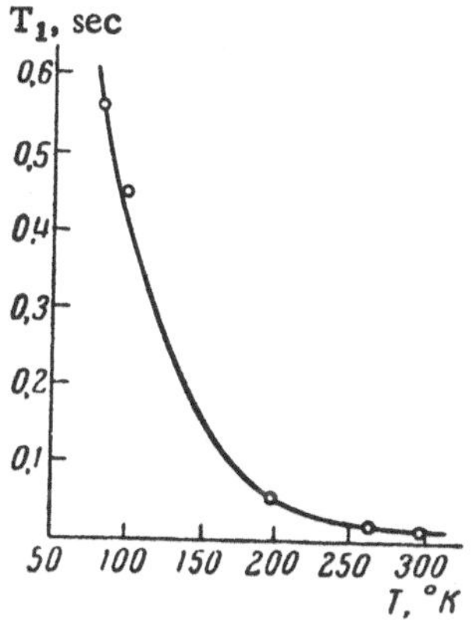

Fig. 13. Variation of T_1 in p-dichlorobenzene with temperature.

predicting values of N for several systems, with the assumption that decrease of A in a solid solution is due to statistical deformations of the lattice. This assumption was based on the results of measurements of the spin–lattice relaxation time T_1 in a solid solution [66], which show that T_1 is independent of c. At the same time, according to Hirai [73], T_1 is very strongly independent of the temperature (Fig. 13). Therefore, it may be assumed that in the first approximation the lattice dynamics remains unchanged at low c. This is also indicated by the absence of frequency shifts in the molecules of the main substance. It then follows from an analysis of the experimental data that deformation of the lattice around each impurity molecule diminishes with increasing distance from that molecule. It is possible to derive numerical characteristics for this decrease of deformation; with their aid, and on the basis of existing data on the dimensions of the impurity molecules and the molecules of the main substance [58], it is possible to predict the dimensions of the deformation region which arises when one molecule of the main substance is replaced by the impurity. The agreement with experimental data is quite satisfactory. It must be pointed out that the calculations can be completed only with the hypothesis that the size of the atoms involved in the intermolecular contacts is proportional to the variations of the corresponding interatomic distances. The interesting question of the meaning of this hypothesis and of the generality of the proportionality factor used, which was taken as close to unity in the first calculations, requires additional study. For this it will be necessary to examine the difficulties involved in applying to analysis of crystallographic splitting of NQR lines the relationship established in [68] between the relative change of frequency $\Delta\nu/\nu$ and the relative scatter $\Delta r/r$ of the intermolecular distances in the crystal:

$$\Delta\nu/\nu = -3\Delta r/r. \qquad (7)$$

The effects of mechanical and thermal treatments of the specimens on intensity of quadrupole resonance lines have been studied in [8, 66, 68]. Dreyfus [69] observed that the NQR signal was stronger in certain solid solutions than in the corresponding pure specimens; this was explained [68] by the effect of strong overheating of the melt on the dimensions of the mosaic blocks in the crystal.

Investigations of the influence of the pressure applied to the specimen on NQR frequencies [74-76] are important. Certain conclusions drawn from these experiments were discussed in Part I, Section 6. In studies of the effect of temperature on NQR frequencies, the variations of lattice constants with temperature must be taken into account. An extensive field for investigations of intermolecular interaction in crystals is also offered by frequency shifts under the influence of pressure.

Influence of Thermal Vibrations on NQR. Because of the exceptionally strong influence of temperature of the specimen on NQR frequency (see Part I, Section 6), quadrupole resonance is a fairly sensitive indicator of thermal vibrations in a crystal. The work of Dodgen, Ragle, and Anderson [77- 79] and of Skripov and Grechishkin [80] is the most significant with respect to applications of this property of NQR. The influence of reorientation about the Cl–Cl axis on the breadth of the NQR line in 1,2-dichloroethane was studied [77]. Increase of the amplitude and frequency of orientation resulted in an eightfold broadening of the line when the temperature was raised from 77°K to 154°K. The observed anomalous frequency shift was attributed by Ragle to transition to random reorientation of 1,2-dichloroethane molecules at a frequency $>10^8$ sec^{-1}. It must be emphasized that the occurrence of such motion must indicate a phase transition into the gaseous crystalline state; i.e., it must be accompanied by an abrupt change in the breadth of the proton resonance line. Since Ragle did not detect any such change, his interpretation of the temperature–frequency relationship for 1,2-dichloroethane must be regarded as debatable. However, a similar interpretation for changes in the NQR spectrum of 1,2-dibromoethane is offered by Dodgen and Anderson [79], who attribute the transition from a two-component to a one-component spectrum at −24° to the occurrence of rapid reorientation of the molecules about the Br–Br axis. Skripov and Grechishkin [80] investigated the average life of the rotational oscillation quanta in a number of compounds [80]. They also considered the origin of the partial rotation of hexachloroethane, trichloroacetic acid, and chloral hydrate molecules; the rotating molecule was considered as a defect in the crystal lattice.

The combined influence of lattice disorder and thermal vibrations makes it impossible to observe NQR in many cases, because of excessive broadening of the resonance line. Unsuccessful attempts to detect quadrupole resonance have been reported by several authors. The summary of these data in Table 2 is certainly not complete. An examination of this table suggests that, for many of the compounds studied, the NQR lines are very much broadened owing to statistically random orientation of molecules in the crystals [81, 82], while, in other cases, thermal motions may be an obstacle to detection of the signal. Obviously, there is a risk that any estimates of line breadths based on a total absence of an NQR signal may be inadequate or quite erroneous. The problem of the causes of the weakness of NQR signals for many substances can be finally solved only when improvements in the apparatus will make it possible to detect quadrupole resonance in at least some of these substances.

The authors are sincerely grateful to A. I. Kitaigorodskii and Yu. T. Struchkov for valuable advice and constant attention.

LITERATURE CITED

1. E. R. Andrew, Nuclear Magnetic Resonance [Russian translation] (IL, Moscow, 1957).
2. H. Dehmelt, Am. J. Phys. 22, 110 (1954).
3. I. Shpigel', M. Raizer, and E. Myae, Zhur. Tekh. Fiz. 27, 387 (1957).
4. M. Buyle-Bodin, J. Phys. Radium 20, 159A (1959).
5. N. Hopkins, Rev. Sci. Instr. 20, 401 (1949).
6. R. Livingston, Ann. N. Y. Acad. Sci. 55, 800 (1952).
7. H. Gutowsky, Rev. Sci. Instr. 24, 678 (1953).
8. T. Wang, Phys. Rev. 99, 566 (1955).
9. E. I. Fedin and G. K. Semin, Radiotekhnika i Élektronika 4, 127 (1959).
10. R. Pound and W. Knight, Rev. Sci. Instr. 21, 219 (1950).
11. R. Pound and G. Watkins, Progr. Nucl. Phys. 2, 21 (1952).
12. J. Cohen and W. Tanttila, Am. J. Phys. 26, 381 (1958).
13. D. Jennings and W. Tanttila, Rev. Sci. Instr. 30, 137 (1959).
14. R. Pound and R. Freeman, Rev. Sci. Instr. 31, 204 (1960).
15. L. Buss and L. Bogart, Rev. Sci. Instr. 31, 204 (1960).
16. C. Dean and M. Pollak, Rev. Sci. Instr. 29, 630 (1958).
17. T. Das and E. Hahn, Nuclear Quadrupole Resonance Spectroscopy (New York, 1958).
18. C. Dean, Rev. Sci. Instr. 29, 1047 (1958).
19. I. A. Safin, Proceedings of Conference on Paramagnetic Resonance [in Russian] (Kazan', 1959).
20. H. Dehmelt, Z. f. Phys. 130, 385 (1951).
21. S. Kojima, K. Tsukada, A. Shimacuhi, and Y. Hinaga, J. Phys. Soc. Japan 9, 795 (1954).
22. W. Laurita and W. Koski, J. Am. Chem. Soc. 81, 3179 (1958).
23. K. Shimomura, J. Sci. Hiroshima Univ. 17A, 383 (1954).
24. S. Kojima, A. Shimacuhi, S. Haginaro, and I. Abe, J. Phys. Soc. Japan 10, 931 (1955).
25. A. A. Kharkevich, Nonlinear and Parametric Phenomena in Radio Technology [in Russian] (GTTI, 1956), Section 38.
26. É. I. Fedin and Yu. S. Konstantinov, Pribory i Tekh. Éksp. 2, 27 (1959).
27. Yu. S. Konstantinov, Pribory i Tekh. Eksp. 6, 134 (1959).
28. M. Bloom, E. Hahn, and B. Herzog, Phys. Rev. 97, 1699 (1955).
29. J. Buchta, Rev. Sci. Instr. 29, 55 (1958).
30. S. Kojima and K. Tsukada, J. Phys. Soc. Japan 10, 591 (1955).

31. C. Dean and R. Pound, J. Chem. Phys. 20, 195 (1952).
32. K. Shimomura, J. Phys. Soc. Japan 14, 235 (1959).
33. J. Chem. Phys. 19, 803 (1951).
34. H. Meal, J. Chem. Phys. 24, 1011 (1956).
35. F. Adrian, J. Chem. Phys. 29, 1381 (1958).
36. V. Morino, T. Chiba, K. Shimomura, and M. Toyama, J. Phys. Soc. Japan 13, 869 (1958).
37. V. Morina and M. Toyama, J. Phys. Soc. Japan 15, 288 (1960).
38. C. Dean, M. Pollak, B. Craven, and G. Jeffrey, Acta crystallogr. 11, 710 (1958).
39. H. Zeldes and R. Livingston, J. Chem. Phys. 26, 1102 (1957).
40. T. Chiba, J. Phys. Soc. Japan 13, 860 (1958).
41. J. Abe, J. Phys. Soc. Japan 13, 918 (1958).
42. K. Shimomura and N. Inoue, J. Phys. Soc. Japan 14, 86 (1959).
43. P. Casabella and R. Bernes, J. Chem. Phys. 30, 1393 (1959).
44. M. Dewar and E. Lucken, J. Chem. Soc. (London) 2653 (1958).
45. M. Dewar and E. Lucken, J. Chem. Soc. (London) 426 (1958).
46. H. Meal, J. Am. Chem. Soc. 74, 6121 (1952).
47. L. Hammett, Physical Organic Chemistry (New York, 1940), Chapter 7.
48. P. Bray, J. Chem. Phys. 22, 1787 (1954).
49. P. Bray, J. Chem. Phys. 22, 2023 (1954).
50. P. Casabella and P. Bray, J. Chem. Phys. 25, 1280 (1956).
51. P. Bray, S. Moskowitz, H. Hooper, R. Barnes, and S. Segel, J. Chem. Phys. 28, 99 (1958).
52. P. Casabella and P. Bray, J. Chem. Phys. 29, 1105 (1958).
53. P. Bray and R. Barnes, J. Chem. Phys. 27, 551 (1957).
54. R. Livingston, J. Phys. Chem. 57, 496 (1953).
55. J. Duchesne and A. Monfils, J. Chem. Phys. 2, 562 (1954).
56. P. Bray, R. Barnes, and R. Bersohn, J. Chem. Phys. 25, 813 (1956).
57. Yu. T. Struchkov and I. N. Strel'tsova, Zhur. Strukt. Khim. 2, 3 (1961).
58. A. I. Kitaigorodskii, Organic Crystal Chemistry [in Russian] (Izd. AN SSSR, Moscow, 1955).
59. J. Duchesne, Arch. Sci. 11, Fasc. spec. 310 (1958).
60. A. I. Kitaigorodskii and É. I. Fedin, Doklady Akad. Nauk SSSR 130, 1005 (1960).
61. I. Randall, W. Molton, and W. Ard, J. Chem. Phys. 31, 730 (1959).
62. J. Duchesne and A. Monfils, Compt. rend. 238, 1801 (1954).
63. C. Dean, J. Chem. Phys. 23, 1734 (1955).
64. S. Segel and M. Lutz, Phys. Rev. 1183A (1955).
65. A. Monfils and D. Grosjean, Physika 22, No. 6 (1956).
66. D. Woessner and H. Gutowsky, J. Chem. Phys. 27, 1072 (1957).
67. S. Kojima, S. Ogawa, M. Minematzu, and M. Tanaka, J. Phys. Soc. Japan 13, 446 (1958).
68. É. I. Fedin and A. I. Kitaigorodskii, Kristallografiya 6, 3 (1961).
69. B. Dreyfus, Ann. de Physique 3, 638 (1958).
70. R. Baer and C. Dean, J. Chem. Phys. 31, 1690 (1959).
71. G. Adler, D. Ballantine, and B. Baysal, International Symposium on Macromolecular Chemistry [Russian translation] (Izd. AN SSSR, Moscow, 1960) Vol. 2, p. 396.
72. É. I. Fedin, Zhur. Strukt. Khim. 2, 2 (1961).
73. A. Hirai, J. Phys. Soc. Japan 15, 201 (1960).
74. T. Kushida, D. Benedek, and N. Bloembergen, Phys. Rev. 104, 1364 (1956).
75. H. Gutowsky and Q. Williams, Phys. Rev. 105, 464 (1957).
76. A. Globa, Compt. rend. 248, 1983 (1959).
77. J. Ragle, J. Phys. Chem. 63, 1395 (1959).
78. H. Dodgen and J. Ragle, J. Chem. Phys. 25, 376 (1956).
79. H. Dodgen and R. Anderson, J. Chem. Phys. 31, 851 (1959).
80. V. S. Grechishkin and F. I. Skripov, Doklady Akad. Nauk SSSR 126, 1229 (1959).
81. T. L. Khotsyanova, Candidate's Dissertation [in Russian] (Institute of Heteroorganic Compounds, AN SSSR, Moscow, 1958).
82. I. N. Strel'tsova and Yu. T. Struchkov, Izvest. Akad. Nauk SSSR, Ser. Khim. 4 (1961).

83. H. Krüger and U. Meyer-Berkhout, Z. f. Phys. 132, 171 (1952).
84. H. Dehmelt, H. Robinson, and W. Gordy, Phys. Rev. 93, 480 (1954).
85. T. Das, J. Chem. Phys. 27, No. 1 (1957).
86. H. Dehmelt, Z. f. Phys. 134, 642 (1953).
87. R. Haering and G. Volkoff, Canad. J. Phys. 34, 577 (1956).
88. H. Dehmelt, Phys. Rev. 92, 1240 (1953).
89. R. Barnes, S. Segel, P. Bray, and P. Casabella, J. Chem. Phys. 26, 1345 (1957).
90. T. O'Konsky and P. Flautt, J. Chem. Phys. 27, 815 (1957).
91. G. Watkins and R. Pound, Phys. Rev. 85, 1062 (1952).
92. M. Minematsu, J. Phys. Soc. Japan 14, 1030 (1959).
93. R. Cotts and W. Knight, Phys. Rev. 96, 1285 (1954).
94. P. Casabella and P. Bray, J. Chem. Phys. 28, 1182 (1958).
95. A. Schwalow, J. Chem. Phys. 22, 1211 (1954).
96. S. Kojima and M. Minematsu, J. Phys. Soc. Japan 15, 355 (1960).
97. R. Barnes and P. Bray, J. Chem. Phys. 23, 407 (1955).
98. H. Krüger and U. Meyer-Berkont, Z. f. Phys. 132, 221 (1952).
99. P. Bray, C. O'Keefe, and R. Barnes, J. Chem. Phys. 25, 792 (1956).
100. S. Ogawa, J. Phys. Soc. Japan 13, 618 (1958).
101. S. Kojima, J. Phys. Soc. Japan 9, 805 (1954).
102. N. Negita, P. Casabella, and P. Bray, J. Chem. Phys. 32, 314 (1960).
103. R. Barnes and P. Bray, J. Chem. Phys. 23, 1177 (1955).
104. S. Ogawa, J. Phys. Soc. Japan 13, 618 (1958).
105. H. Dehmelt, Z. f. Phys. 130, 385 (1951).
106. H. Robinson, H. Dehmelt, and W. Gordy, Phys. Rev. 89, 1305 (1953).
107. H. Robinson, Phys. Rev. 100, 1731 (1955).
108. H. Dehmelt, Phys. Rev. 91, 313 (1953).
109. H. Gutowsky and D. McCall, J. Chem. Phys. 32, 549 (1960).
110. H. Negita and Z. Hirano, Bull. Chem. Soc. Japan 31, 660 (1958).
111. R. Livingston, Phys. Rev. 82, 289 (1951).
112. R. Livingston, Record. Chem. Progr. 20, 173 (1959).
113. H. Allen, J. Phys. Chem. 57, 501 (1953).
114. P. Bray, J. Chem. Phys. 23, 703 (1955).
115. J. Graybeal and C. Cornwell, J. Phys. Chem. 62, 483 (1958).
116. E. Lucken, J. Chem. Soc. (London) 2954 (1959).
117. H. Dehmelt, Z. f. Phys. 129, 401 (1951).
118. D. McCall and H. Gutowsky, J. Chem. Phys. 21, 1300 (1953).
119. T. Weatherley and Q. Williams, J. Chem. Phys. 22 (1954).
120. K. Torizuka, J. Phys. Soc. Japan 9, 645 (1954).
121. H. Hooper and P. Bray, J. Chem. Phys. 30, 957 (1959).
122. P. Bray and P. Ring, J. Chem. Phys. 21, 2226 (1953).
123. S. Segel, R. Barnes, and P. Bray, J. Chem. Phys. 25, 1286 (1956).
124. T. Weatherley, P. Davidson, and Q. Williams, J. Chem. Phys. 21, 761 (1953).
125. T. Weatherley and Q. Williams, J. Chem. Phys. 21, 2073 (1953).
126. P. Bray and D. Esteva, J. Chem. Phys. 22, 570 (1954).
127. P. Bray, J. Chem. Phys. 23, 220 (1955).
128. A. Monfils, Compt. rend. 241, 561 (1955).
129. É. I. Fedin, Zhur. Strukt. Khim. 1, 124 (1960).
130. A. Monfils and J. Duchesne, J. Chem. Phys. 22, 1275 (1954).
131. S. Kojima, Bull. Chem. Soc. Japan 28, 271 (1955).
132. Q. Williams and T. Weatherley, J. Chem. Phys. 22, 572 (1954).
133. R. Brown, J. Chem. Phys. 32, 116 (1960).
134. K. Kojima and S. Saito, J. Chem. Phys. 30, 560 (1959).
135. R. Barnes and S. Segel, Phys. Rev. Letters 3, 462 (1959).
136. R. Barnes and R. Engardt, J. Chem. Phys. 29, 248 (1958).

137. N. Negita and S. Saton, Bull. Chem. Soc. Japan 29, 426 (1956).
138. K. Torizuka, J. Phys. Soc. Japan 11, 84 (1956).
139. J. Ragle, J. Chem. Phys. 32, 403 (1960).
140. V. S. Grechishkin, Priroda 8, 85 (1959).
141. S. Kojima, K. Tsukada, S. Ogawa, and A. Shimauchi, J. Chem. Phys. 23, 1963 (1955).
142. C. Cornwell and R. Jamasaki, J. Chem. Phys. 27, 1060 (1957).
143. Y. Kurita, P. Nakamura, and S. Hayakawa, J. Chem. Soc. Japan. 79, 1093 (1958).
144. R. Nakamura, Y. Uehara, Y. Kurita, and M. Kubo, J. Chem. Phys. 31, 1433 (1959).
145. P. Barnes and S. Segel, J. Phys. 25, 180 (1956).
146. I. Hatton and B. Rollin, Trans. Faraday Soc. 50, 358 (1954).
147. H. Zeldes and R. Livingston, J. Chem. Phys. 21, 1418 (1953).
148. S. Kojima, K. Tsukada, S. Ogawa, and A. Shimauchi, J. Chem. Phys. 21, 1415 (1953).
149. G. K. Semin, Zhur. Strukt. Khim. (in press).
150. J. Ludwig, J. Chem. Phys. 25, 950 (1956).
151. D. Alderdice, R. Brown, and T. Iredale, J. Chem. Phys. 32, 314 (1960).
152. P. Bray, J. Chem. Phys. 22, 950 (1954).
153. H. Robinson, H. Dehmelt, and W. Gordy, J. Chem. Phys. 22, 511 (1954).
154. K. Shimomura, J. Chem. Phys. 22, 1944 (1954).
155. H. Dehmelt, Z. f. Phys. 129, 401 (1951).
156. R. Barnes and S. Segel, J. Chem. Phys. 25, 578 (1956).
157. F. Herlach, H. Gränicher, and D. Itschner, Arch. sci. (Geneva) 12, Fasc. spec. 182 (1959).
158. R. Jamasaki and C. Cornwell, J. Chem. Phys. 30, 1265 (1959).
159. I. Voitländer and R. Longino, Naturwiss. 46, 664 (1959).
160. R. Livingston and H. Zeldes, J. Chem. Phys. 26, 351 (1957).